P9-CCQ-311

# BONES

# BONES

## *Inside and Out*

Roy A. Meals, MD

W. W. NORTON & COMPANY
*Independent Publishers Since 1923*

Printed in the United States of America
First Edition

For information about permission to reproduce selections from this book, write to Permissions, W. W. Norton & Company, Inc., 500 Fifth Avenue, New York, NY 10110

For information about special discounts for bulk purchases, please contact W. W. Norton Special Sales at specialsales@wwnorton.com or 800-233-4830

Manufacturing by Lake Book Manufacturing
Book design by Lovedog Studio
Production manager: Lauren Abbate

Names: Meals, Roy A., author.
Title: Bones : inside and out / Roy A. Meals, MD.
Description: First edition. | New York, NY : W. W. Norton & Company, [2020] | Includes bibliographical references and index.
Identifiers: LCCN 2020013491 | ISBN 9781324005322 (hardcover) | ISBN 9781324005339 (epub)
Subjects: LCSH: Bones.
Classification: LCC QL821 .M525 2020 | DDC 599.9/47—dc23
LC record available at https://lccn.loc.gov/2020013491

W. W. Norton & Company, Inc., 500 Fifth Avenue, New York, N.Y. 10110
www.wwnorton.com

W. W. Norton & Company Ltd., 15 Carlisle Street, London W1D 3BS

1 2 3 4 5 6 7 8 9 0

*To Susan,*

*whose love, support, impeccable taste,*

*and gentle guidance greatly enrich*

*my writing and my life.*

# *Contents*

# *Introduction*

Consider the shortcomings of everyday building materials. Mud mushes. When it dries out, it crumbles. Limestone, granite, concrete, brick, and china don't crumble, but they are brittle and their weight and bulk limit their usefulness for building, especially for things that are supposed to move. Metal makes for lighter construction, and if you bend metal a bit, it springs back, which is usually fine. But bend metal some more and it stays bent, which can be bad. Plastics are environmentally unfriendly. Wood is good because it is a bit flexible, easy to join, relatively lightweight, and biodegradable; but it can rot or burn.

Other building materials, used by living organisms, also have their shortcomings. Shells are heavy, which makes it difficult for snails and clams to move quickly. Lobsters are fast, and beetles can even fly, but their thin, lightweight, crunchy exterior support system needs to be vacated and replaced periodically in order for the owner to grow.

All this leads to bone. Not only is bone manufactured on-site, it is also lightweight, durable, and responsive to changing conditions. A bridge made of steel cannot double its length or its carrying capacity, but bone both grows and responds to stresses. Furthermore, bone mends itself. A shattered brick or a broken spoon, be it made of metal, plastic, or wood, cannot do that. Not only is bone the world's best structural support, it is also the world's largest import-export bank, a repository of vital elements—especially calcium—on which our lives depend.

Despite all of its marvels, hardly anybody has ever seen or wants to see living bone, especially their own. So bone, regardless of its superlative features, lives in seclusion and does not get the respect it deserves. For example, what image pops into your mind when you think of bone? Is it a Georgia O'Keeffe painting of a cow skull toasting in the sun? That desert depiction—dry, white, timeless—does not begin to do bone justice. Furthermore, annoyance and even disdain for bone arise when one tries to get that last morsel of meat off a rack of lamb or a beef T-bone. Rushing toward dessert, you probably have never paused to marvel at that ring of bone centered in your ham steak, to understand why drumsticks are thicker at the ends than in the middle, or to ponder why some fish bones are rubbery and wishbones are brittle. Cynics who remain unconvinced that bone is the best building material may even ask, "If bones are all that great, why don't snails have any? Or bees?" I will answer these questions and many more as the story of bone unfolds.

Although bone is ubiquitous and versatile, it is rarely seen in its living state, and hence it is a bit mysterious. Then after this marvelous mystery material has finished serving and protecting its original owner, it reveals itself in myriad places and for myriad purposes, sometimes hundreds of millions of years later. Bone has much to teach us about Earth's history and the course of animal life on the planet. Additionally, from the dawn of civilization, humankind has repurposed bone to serve and protect, even to amuse and inspire. Bone's durability and ubiquity make its revealed state as interesting as its concealed state. By the end of the book, you'll be convinced that it is the world's best building material.

*Part One*

# BONE CONCEALED

*Chapter 1*

# BONE'S UNIQUE COMPOSITION AND VARIED STRUCTURE

THE PROMINENT GREEK PHYSICIAN AND philosopher Galen wrote that bone, because of its pale color, was made of sperm. One thousand years later, Avicenna, a Persian astronomer, physician, and prolific writer, thought that bone was made of earth because it was cold and dry. Now another one thousand years later, different notions prevail. Avicenna did, however, note that the best way to understand the skeleton was to separate it from the rest of the body, which is still good advice.

To get to know bone, we should separate it down to its organic chemistry. Five carbon atoms can combine with one each of oxygen and nitrogen along with several of hydrogen to make proline, an amino acid. Amino acids are the essential building blocks of life, and humans both make proline from scratch and break it down by digesting proteins. Certain cells then assemble a proline-rich mix of amino acids into a chain to make a molecule of collagen, which is the most common protein in our bodies. The strand at this point faintly resembles an ultramicroscopic strand of limp spaghetti. An extra hydrogen and oxygen atom then attach to many of the proline molecules. This causes the chain to kink sharply at regular intervals, and it now looks more like a minuscule piece of corkscrew pasta. Three of these corkscrews nest together to form a collagen molecule, which even at this super-submicroscopic level is strong and stable because the nooks in one strand nest into the crannies of the adjacent ones.

Collagen molecule assembly occurs in several different kinds of cells, including the type that makes bone. These are the osteoblasts (from the Greek words for "bone" and "germinator"). After forming a collagen molecule, the osteoblast pushes this chemical-mechanical wonder through its cell membrane out into the minute space between the cells. In this new location, collagen molecules not only join end to end but also snuggle side to side to make a multistrand fiber. The collagen molecules are so thin that it would take seventeen hours of adding one collagen molecule every second to make a stack as tall as this page is thick. And even though the molecules are much longer than they are thick, it would take 300,000 of them placed end to end to span the space inside this "o."

The collagen fibers are locked together both by mechanical nesting (nook-cranny) and by chemical (sticky pasta-like) bonds. If you consider how strong three tiers of Legos would be if you smeared them with Super Glue before snapping them together, then you begin to understand the strength of collagen fibers. When stretched, these tough amino acid chains are stronger than equally thick filaments of steel.

That is all the chemistry needed for now. Let's take a minute to understand the connection between Limeys, shoe leather, furniture glue, and jiggly gelatin desserts. Remember the hydrogen-oxygen appendages on proline molecules? Vitamin C catalyzes those attachments. Without them, no tight corkscrewing occurs. A deficiency of vitamin C leads to faulty collagen production, which is the cause of scurvy. Symptoms include bleeding gums and easy bruising. Early mariners were at sea for months and endured diets almost entirely devoid of fresh fruits and vegetables. British sailors learned to squeeze fresh limes into their fetid drinking water to mask its taste. By doing so, their collagen production returned to normal, and the sailors discovered serendipitously that a lime a day kept the scurvy away.

Another collagen connection relates to shoe leather, which is tanned cowhide. It is tough because the tanner adds chemicals to the vats that increase the number of weld points among the collagen fibers in the hide. Just the opposite is true for paintball covers, medicine capsules, hide-based furniture glue, gelatin desserts, and gummy bears. They are all made from partially unlinked collagen, which comes from boiled by-products of meat and leather production. This is where the expression "sending a horse to the glue factory" came from. Warning: do not read a description of gelatin manufacture if you want to enjoy your next marshmallow.

Collagen resists stretching. It is the main constituent of tendons, which convert muscle contractions into joint motions, and ligaments, which keep joints properly aligned. Consider standing on your toes. If your Achilles tendons were rubbery, they would stretch like bungee cords when you contracted your calf muscles, and your heels would remain on the floor. This would be bad. You could never jump. Or consider putting your fingertips on your cheek and stretching your fingers away from your palm. If your ligaments were not tough, you could push your fingers away until your fingernails touched the back of your hand. This would be unsightly. Some people naturally have ligaments that are extremely stretchable. These so-called double-jointed contortionists enjoy demonstrating their great flexibility while watching the rest of us wince.

You may have considered the description of collagen irrelevant, since collagen is tough and does not stretch, and we know intuitively that bone is not stretchy. Rather, bone is rigid and resists getting mashed flat. It resists compression (scientific parlance for mashing) because it consists of calcium crystals deposited on a meshwork of—you guessed it—collagen, like plaster on lath.

To prove it for yourself, buy a pack of chicken drumsticks. Filet several of them and soak the bones in vinegar for several weeks. Cook the others, enjoy a chicken dinner, and then bake the leftover bones at 250 degrees Fahrenheit for several hours. The ones soaked in vinegar become rubbery and bendable because the vinegar dissolves the calcium from the collagen meshwork. The baked one becomes brittle and fragile, like sticks of chalk, because the heat destroys the collagen.

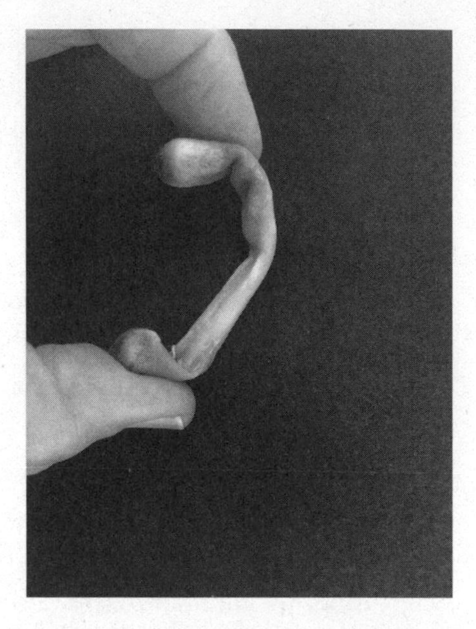

*After soaking in vinegar, this chicken leg bone has lost the rigidity provided by calcium crystals. What remains is the flexible collagen framework.*

Chemistry books note that calcium crystals come in various forms. These include calcium chloride (road deicer), calcium citrate (water softener, diet supplement), calcium carbonate (antacid tablets, chalk, coral, eggshell), calcium sulfate (gypsum, plaster of paris), and calcium hydroxide (slaked lime). If you add a phosphorus compound to calcium hydroxide under the right conditions, you get hydroxyapatite, which might be a new word for you. It has nothing to do with Hydrox cookies or appetite, even for marshmallows or other gelatin delectables. Rather, hydroxyapatite is the principal calcium crystal of bone. People might consider you a bit weird if you tossed the word out at a cocktail party, but it is what our bone crystals are called, and hydroxyapatite is holding you off the floor, so let's learn a little about it.

In the 1780s, a German mineralogist recognized the apatite crystal as a distinct species, whereas previously it had been confused with other minerals or repeatedly identified as new ones. He named it for its deceptive nature: *apatit* (in German), from the Greek *apatē*, meaning "deceit." The mineral exists in various forms. With a water ion attached, it becomes hydroxyapatite.

Particularly if you are watching your weight, understand that bone constitutes about 15 percent of our body mass and that bone is about one-third collagen and two-thirds calcium-phosphorus crystals. Accordingly, a 160-pound person possesses 24 pounds of bone (8 pounds of collagen and 16 pounds of hydroxyapatite). That's enough to fill a carry-on roller bag, not that you would want to try to slip it past TSA, but at least you have a mental image now of one's bone mass.

Imagine some osteoblasts in a bread pan floating around in a nutritious broth of water and oxygen. The osteoblasts follow their genetically programmed urge to manufacture and secrete collagen and hydroxyapatite molecules. Voilà, the calcium crystals deposit themselves on the meshwork of collagen, and we get bone. The osteoblasts essentially seal themselves in cocoons of bone, where they become osteocytes—mature bone cells. They maintain the bone's structure but do not contribute much to further construction or to any destruction. Various chemical messengers, mainly hormones from the pituitary, thyroid, and testes or ovaries, affect the vigor with which osteoblasts produce bone. Other nearby cells produce chemical messengers known as growth factors, which also consist of amino acid chains. Several of these growth factors can whip the osteoblasts into a bone-forming frenzy and even convert several other types of cells into bone-forming cells if necessary.

When enough osteoblasts have done their job encasing themselves next to one another in cocoons of collagen-reinforced hydroxyapatite, the nutritional broth in the bread pan turns rock solid. The lump would be about the same density and strength as an adobe brick. Can you imagine our ancestors successfully avoiding pursuit by lions if their bones were made of adobe? Ah, but consider further, the lions would have similarly formed bones—so it would have been a boring, slow-motion chase. Of course, that is not what happened evolutionarily. To understand what did happen, we need to consider some mechanical principles, which will explain why most flat bones (for example, skull and breastbone) have two layers of compact bone sandwiching a spongy interior and why the long bones in the arms and legs are cylindrical, just like tubes in a bicycle frame.

First, consider the thin flat bones. The skull protects the brain, and the breastbone and ribs protect the heart and lungs from direct blows. The inner and outer surfaces of these plates are hard, dense, and smooth, and they resist bending and puncture. The spongy interior is rigid (think frozen sponge), lightweight, and adds strength—identical to the structure of corrugated cardboard.

Now, consider the tubular bones. To appreciate the beauty of their structure, first think about a wooden board that is 10 feet long, 18 inches wide, and 2 inches thick. We could use it to span an 8-foot-wide chasm and safely walk across. A little springy perhaps, but we would make it. To eliminate the springiness, we could turn the plank up sideways and tiptoe across on the 2-inch-wide edge. The bridge would be a lot narrower but far more rigid. In these two arrangements, the board does not change its dimensions or physical properties, but when turned on edge, it has 18 inches of material arranged vertically to resist bending rather than just 2 inches.

That is why the floor joists in a wood frame house are set on edge. Otherwise the floor would be a trampoline. You could make floor joists really thick and lay them flat; but by the time you got them thick enough so that they were not springy, the floor would be so heavy and expensive that the entire project would likely collapse both mechanically and financially.

How do engineers make joists, beams, and girders maximally efficient? In other words, how do they achieve the largest effect while using the least amount of material and effort? They choose beam designs that when viewed end on have the shape of a capital I—the "I-beam." We could discuss the principle of the I-beam using formulas that contain Greek letters, but here is a pain-free overview. The portions of a beam or joist that contribute the most to its stiffness are the portions that are near its edges. For example, you could start with a rectangular beam and remove much of the material that is on its sides. The beam would maintain most of its original strength but with greatly reduced weight and cost.

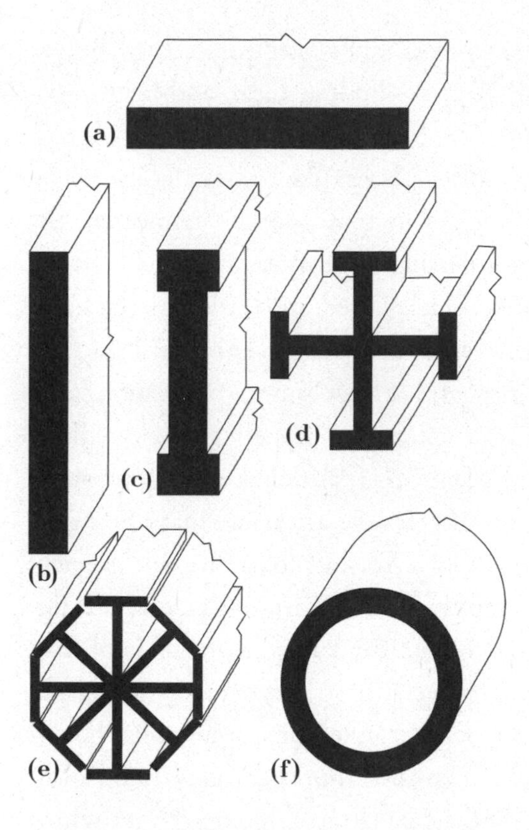


*These forms all contain the same amount of material when their lengths are equal. The flat plank (a) is springy against vertical forces. Turned on edge (b), the form can resist vertical bending. An I-beam (c) is even stronger. Although they are both imaginary, the crossed I-beams (d) would resist both vertical and horizontal bending, and a multiple I-beam structure (e) would effectively resist bending from many directions. The cylinder (f) effectively resists bending from all directions and mimics the structure of bone.*

The good news about an I-beam is that it efficiently resists bending from top-down forces. The bad news is that it is not particularly strong in resisting side-to-side forces or twisting forces. A beam equally resistant to forces from the top *and* side would look something like a skinny iron cross. This beam, however, would still be weak when resisting forces applied obliquely from the two, five, eight, or eleven o'clock positions.

To resist forces coming from any direction, the beam would have to be a composite of numerous I-beams arranged in a circle. If the outer flanges of each portion of the composite beam became joined together, the middle portions could be removed without greatly weakening the structure. What's left? A cylinder, which can resist bending and twisting forces from *all* directions. Its hollow interior spares weight and material, and an identically dimensioned solid rod would not greatly add to the structure's stiffness. That is the beauty of bicycle frames, ski poles, and—you guessed it—bones. Our long bones are basically tubes—hollow, light, and bend resistant in all directions.

Note, too, that most cylindrical bones flare at the ends, where they are covered with cartilage, which is another connective tissue, this time formed by interspersing large molecules on a collagen meshwork. In the case of bone, the "plaster" molecules are the hard, compression-resistant hydroxyapatite crystals. For cartilage, the clinging molecules are springy and attract water. These molecules' spongelike features bestow on cartilage its slippery property, allowing the jointed bone ends to glide nearly frictionless over one another.

The structure and function of cartilage are as fascinating as they are for bone, but cartilage will have to wait for a book of its own. For bone lovers, all we need to know is that cartilage is soft and slippery compared with compact bone. The flares on the ends of long bones protect the delicate cartilage in two ways. First, the flaring increases the area of contact between the ends of adjacent bones, thereby reducing the pressure in any one spot that the cartilage must endure. Second, the flare consists mostly of spongy bone, which is a bit springy and provides a cushioning effect for the pressure-sensitive cartilage.

You have probably noticed that the central cavity inside a long bone's hard, dense cylinder is not entirely empty. This leads us to the nature and purpose of both types of bone, compact and spongy—hard on the outside, springy on the inside, like a Tootsie Pop or a loaf of crusty French bread. Bone's outer surface is mechanically sound and does our heavy lifting. The spongy, central-cavity bone adds some strength and support to the hard surface, especially near the ends.

Filling the spaces created by the spongy bone are two types of marrow cells—red and yellow. Red prevails in early life and is present particularly at the ends of our cylindrical bones. Red marrow has a rich blood supply and is responsible for making new blood

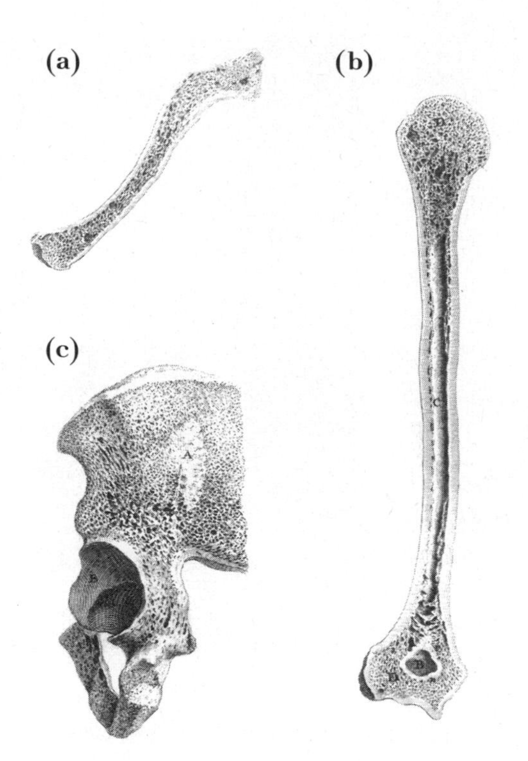


*In his 1733 atlas,* Osteographia, *William Cheselden described this figure showing a collar bone (a), an upper arm bone (b), and a pelvis (c). "Several bones sawed through to shew their inward spongy texture, the cells not being perfectly clear'd of the marrow that dried in them." Bone's spongy interior greatly reduces its weight and enhances its strength, particularly near the ends.* William Cheselden, *Osteographia, or the Anatomy of the Bones* (London: W. Bowyer, 1733).

cells—roughly 500 billion a day. Yellow marrow is mostly gelatinous fat, and with increasing age it constitutes a greater proportion of bone's interior. Some gourmands find yellow marrow from beef bones delicious and will scrape, chew, crack, and even slurp them to get it. This is what Henry David Thoreau meant when he went to the woods "to live deep and suck all the marrow out of life." That is pretty much what I do with Tootsie Pops.

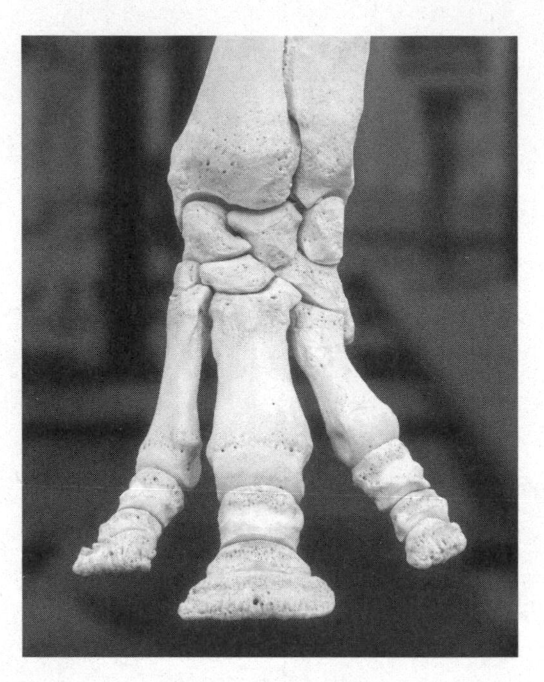

*The forelimb skeleton of this white rhinoceros demonstrates multiple small openings through which blood vessels access each bone's spongy interior and provide nourishment.* SKELETONS: MUSEUM OF OSTEOLOGY IN OKLAHOMA CITY, OKLAHOMA.

Some birds have entirely marrowless thigh and arm bones, which are key parts of their breathing system. These cavities momentarily store inhaled air before it passes through the lungs and back out the mouth. Some dinosaurs had similar hollow bones that also likely aided respiration. These hollow bones in ancient dinosaurs and in modern birds strengthen the evidence that birds developed from those ancient reptiles.

Inquiring minds may now be wondering, How does blood get through a bone's dense cylinder to nourish its spongy interior? A hole passing directly through the bone—one large enough to accommodate sizable blood vessels—would cause problems. Such an opening would weaken the cylinder and greatly reduce its ability to resist bending and twisting forces. Fractures would easily occur. Instead, multiple pinhole tunnels take long, diagonal paths through the cylinder's wall. Each one contains a minuscule artery and vein. Some bones have more of these nutrient passageways than others. The hip bone and one bone in each the wrist and ankle are noted for not having any passages along major portions. This makes healing fractures in those regions problematic because the supply lines for delivering the building materials are limited.

Curious types may also be pondering what the purpose of bone is in the first place. Consider the motto of the Los Angeles Police Department: "To Protect and Serve." From an orthopedic surgeon's point of view, the cops have the order wrong. We think service, then protection. Neurosurgeons and cardiologists might feel differently. Your skull protects your brain, and your ribs and breastbone protect other assorted innards, but the really great bones—the ones whose well-being orthopedists oversee—are the service providers: spine, pelvis, limbs. Each bone's unique shape allows it to provide service and, in some instances, protection as well.

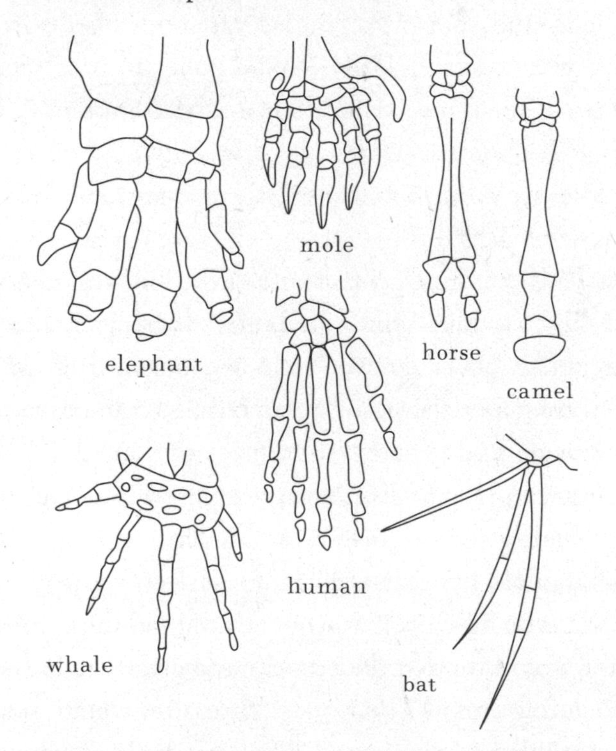


*Modifications of basic skeletal structures render the forelimbs of various animals efficient for specialized activities that include weight-bearing, digging, running, swimming, and flying. The human hand, although not particularly good at any of these activities, has one major advantage—the ability to grasp tools.*

Size aside, there are remarkable similarities between the shapes of analogous bones in most mammals, birds, and even dinosaurs. Take, for example, an elephant's shin bone and a chicken's corresponding drumstick bone. Each is relatively narrow in the middle and flares out at both ends to contribute to the knee and ankle joints. The broad ends serve to widely disperse weight-bearing forces and also provide adequate surface area for attachment of ligaments, which prevent the joints from flopping around.

If someone were to ask you how many bones there are in the human body, please do not blurt out "206." Though 206 is the widely accepted number, in reality the answer is complicated. Consider first that humans vary from one another in facial features, hair color, height, and shoe size. Similar variations also exist beneath the skin. Nerves, tendons, arteries, and bones all have their unique arrangements; their precise location and size inside *my* body say little about their precise arrangement in yours. To solve the bone-count riddle, we have to address five key questions: Who? What? When? Where? Why?

First, *who* is counting? A paleontologist brushing sand off long-buried fossils may miss some tiny bones. These include sesamoids, which are small bones embedded in tendons and found adjacent to joints throughout the body. Sesamoids, so named because they resemble sesame seeds (though in humans they are larger, more like the size of capers), help to distribute pressure evenly as we grip objects with our hands and bear weight on our feet. Some people do not have any sesamoid bones in their hands or feet, yet they manage as well as those who have 20. Consequently, we call these bones "accessories." You may want to include at least some of them in your count.

Next, *what* counts as a bone? The kneecap is a giant sesamoid. It is always included in the favored 206. So is a pea-sized wrist bone. Most people have 24 ribs, 12 on each side of the chest, but others have 26—and they don't get extra credit. Three tiny bones in each ear count toward our total, but sesamoids in the feet are left out, as are bean-sized accessory bones around the hip, knee, and ankle.

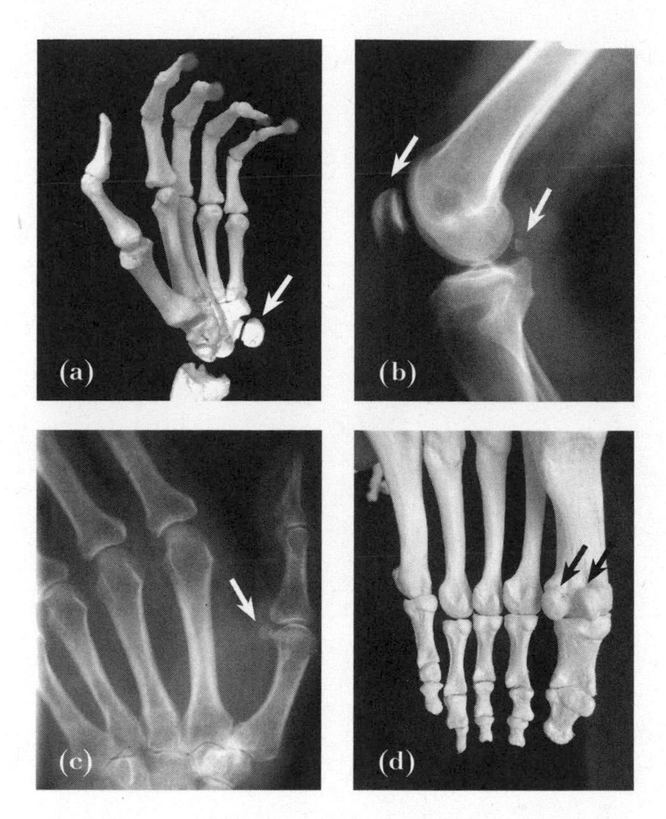


*The sesamoid bones in the heel of the hand (a) and on the front of the knee (b) are included in the classic total-body bone count of 206; but ones on the back of the knee, in the thumb (c), and in the ball of the foot (d) are not included.*

*When* do we count? Babies are born with around 270 bones, and then some slowly fuse together. Initially, the platelike skull bones can move against one another and shift the head's shape to facilitate delivery. Then its a normal occurrence for the skull bones to fuse together in order to protect the brain. In infancy, many bones in the wrists and ankles do not yet contain enough calcium to block X-ray beams and thus don't make shadows during imaging. There are also wrist and anklebones that sometimes for no good reason fuse with neighboring bones, further complicating our count.

*Where* do we look? Different books offer different perspectives on bone count. Depending on its intended audience, one book might

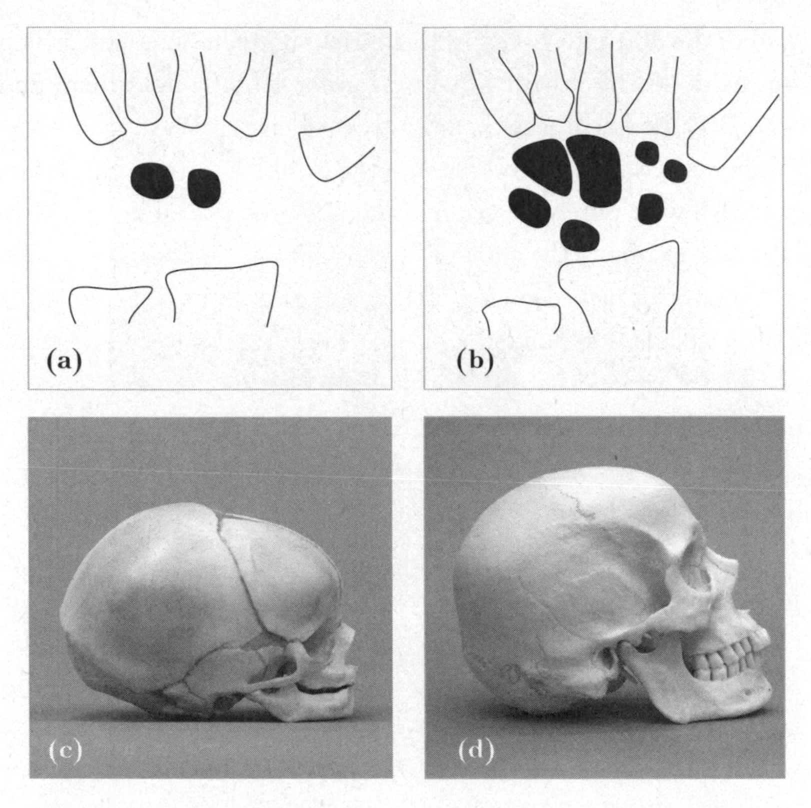


*The number of wrist bones remains constant throughout life. In infancy, however, only two of them contain enough calcium to appear on an X-ray (a). Six years later (b), seven of the eight wrist bones are evident on an identically oriented X-ray. By contrast, the number of bones in the skull diminishes from birth (c) to maturity (d) as multiple small bones fuse together.* (C, D) © BONE CLONES, WWW.BONECLONES.COM.

not show any sesamoid bones, while another will identify any bone that has ever been described.

Finally, *why* bother? For different reasons, the number matters to medical students, surgeons, and paleontologists. Consequently, your best answer to the bone-count question is that nobody really knows—and that you're not going to expose yourself to enough radiation to find out.

All of the 200-plus bones in the human body have names, which helps us to describe them when we cannot actually hold them and point. Because Latin was the original language of Western science, most bones received Latin names, some of which were derived from Greek. All were purely descriptive and self-explanatory if you spoke Latin. For example, the shoulder blade is mostly a flat triangle. An early anatomist picked one up, pondered, and decided it resembled a shovel's blade. He named it *scapula*, Latin for "shovel." The eight wrist bones are another good example of labeling simplicity. First only identified by number, these bones later received Latin names, including *scaphoid*, *lunate*, *triquetrum*, and *pisiform*. The names simply describe the bones' shapes: boat, crescent moon, three corners, and pea, respectively.

Not only do the bones have names that are vaguely descriptive; so do all their bumps, ridges, and nooks. The top of your shoulder blade is the *acromion*, derived from the Greek *acro*, meaning "highest" (as in acropolis—high city), and from *omion*, meaning "shoulder." The tip of your elbow is your *olecranon*. It comes from the Greek *olene* for "elbow" plus *kranion* for "head." The hip joint's deep socket in the pelvis is the *acetabulum*, named for its resemblance to a vinegar cup: *acetum* for "vinegar" plus *abulum* for "container." The bumps on both sides of your ankle? They are *malleoli*, derived from *malleus*, or "hammer." (What were they thinking?)

Doctors today buy into the Greek and Latin naming tradition both necessarily and willingly. Using uniform terminology, medical professionals can understand lectures and journal articles from around the world. Consciously or unconsciously, perhaps doctors also believe that tossing around ancient terminology distinguishes the medical wizards from the uneducated masses. This makes knowledge privileged and hence valuable. When bone wizards gather, we like to regale one another with our experiences treating "the calcaneus," "those condyles," and "that corocoid"; but after reading this book, you won't be fooled. Consider the awe-inspiring *foramen magnum*—the 1-inch-diameter hole at the base of the skull from whence the spinal cord emerges. *Foramen magnum* sounds grand and important, perhaps even a bit magical. It translates, however, as "big hole."

Rather than keeping things murky, songwriter James Weldon Johnson wrote simple lyrics for the spiritual "Dem Bones." Conversely, had the composer been an anatomist, we would likely be singing, "The tibia is connected to the patella, the patella is connected to the femur. . . ."

Such connections are not unique to humans. When I visit a natural history museum, I marvel at the similarity of skeletal structures between widely diverse animals. This is not apparent at the zoo, where an elephant's forefoot is obviously a far cry from a bat's wing. But looking at their exposed bones reveals a similar general organization. Skeletal adaptations between species allow the elephant's foot to bear tremendous weight and the bat's wing to carry its owner aloft, yet their bones reflect a common ancestry.

Nonetheless, some animals have unique and interesting bones that humans do not have. I'll mention five, none of which are numbered among the usual 206 belonging to humans. Each bone demonstrates a special enhancement that contributes to its owner's success. I will move from the familiar to the bizarre.

Some biologists must have finished their Thanksgiving dinner and asked themselves, "We all know how a turkey's wishbone serves us, but what purpose does it serve a turkey?" From a structural perspective, even if you are an expert turkey carver, it has probably escaped you that the wishbone is both collar bones fused together. That, however, does not explain its function, so some enterprising biologists rigged up a system to have starlings fly in a wind tunnel hooked up to an X-ray machine to reveal images of wishbones in action.

As you know, turkey and chicken wishbones are a bit springy. It is the same for starlings. With every powerful downstroke of the bird's

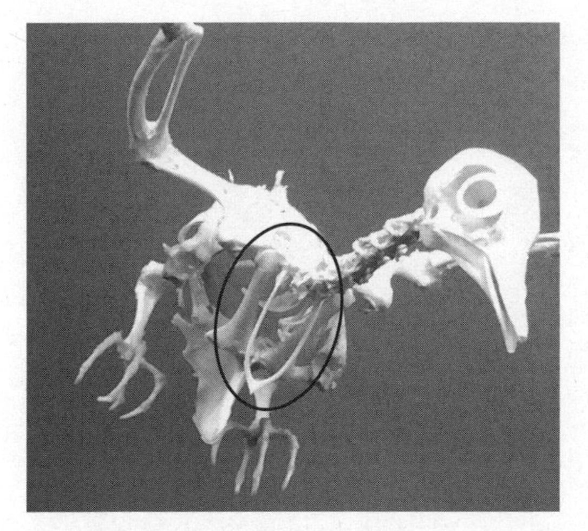

*This pigeon has two bones that humans do not have: a wishbone (circled) and a flat ring in each eye socket.* Museum of Osteology.

wings, the ends of the wishbone move away from each other and absorb energy. Then with the upstroke, which does not need to be as strong, the wishbone springs back to its resting shape and aids the efficiency of flying. Some toucans and owls, however, do not have wishbones and are still capable of strong flight. Conversely, cranes and falcons have rigid wishbones that do not facilitate flight, but their wishbones may help with breathing. Maybe this explains why the turkey became traditional Thanksgiving fare—a roasted toucan or crane would not provide the same after-dinner excitement. Some dinosaurs, including *Tyrannosaurus rex*, also had wishbones, but humans were not around at the time to consider roasting a *T. rex*, much less making a wish on the remains.

The second adaptation also links dinosaurs and birds—hollow bones for breathing. Some dinosaurs took the concept further, for sounding. They had skulls with cavernous bony crests extending back over the tops of their heads. These hollows connected with the nostrils and throat, and the round-trip distance for a breath of air could be as much as 10 feet. Investigators have proposed that this constituted a resonating chamber, which the reptile used to produce and amplify low-pitched calls.

The third special bone is a flat ring that is embedded in the

Parasaurolophus's *long bony crest connected with its nostrils and throat and possibly served as a resonating chamber to amplify its calls.* MICHAEL JABLON, MD.

white of the eye of some birds and other reptiles, including dinosaurs. It surrounds the eyeball and tends to give the skeleton a scholarly look. It probably helps support the shape of the eye. Nobody knows for sure.

The fourth is a *set* of bones, the gastral basket, also possessed by a variety of prehistoric birds and other reptiles, including *T. rex*. Crocodiles and a lizard-like New Zealand creature are the sole owners currently. A gastral basket looks like an oven-rack set of extra ribs at belly level, except that they are not attached to the rest of the skeleton. The gastral basket provided shield-like protection to the owner's soft underbelly and may have contributed to breathing, to flight, or to both.

Nothing is in doubt about the function of the fifth special bone. It supports the penis during erection. It is present in various mam-

*A few ancient birds and some other reptiles, including this extinct marine reptile (with its underside depicted here), possess abdominal ribs. They extend from breastbone to pelvis and are connected to one another but not to the rest of the skeleton.* D. W. Niven.

mals, including dogs, cats, raccoons, walruses, sea lions, and even gorillas and chimps, but not humans. The penis bone allows for prolonged intercourse, which is the necessary strategy for ensuring paternity of offspring when suitable females are encountered infrequently. Its shape varies from rodlike to fantastic. Its size varies from tiny in small monkeys to over 2 feet long in sea lions and walruses. I have seen one that had fractured midshaft and then healed. Poor boy! Females of species that harbor penis bones generally have clitoris bones, although much smaller.

◆◆◆◆

In this chapter, we have covered a lot of territory, beginning with specific arrangements of proline molecules and calcium atoms—chemically, mechanically, and anatomically. Different bones provide unique and varied services for their owners. To do so, they continuously respond to chemical and mechanical influences, and through the ages they have competed with other skeletal support systems. Let's see how bone manages these challenges.

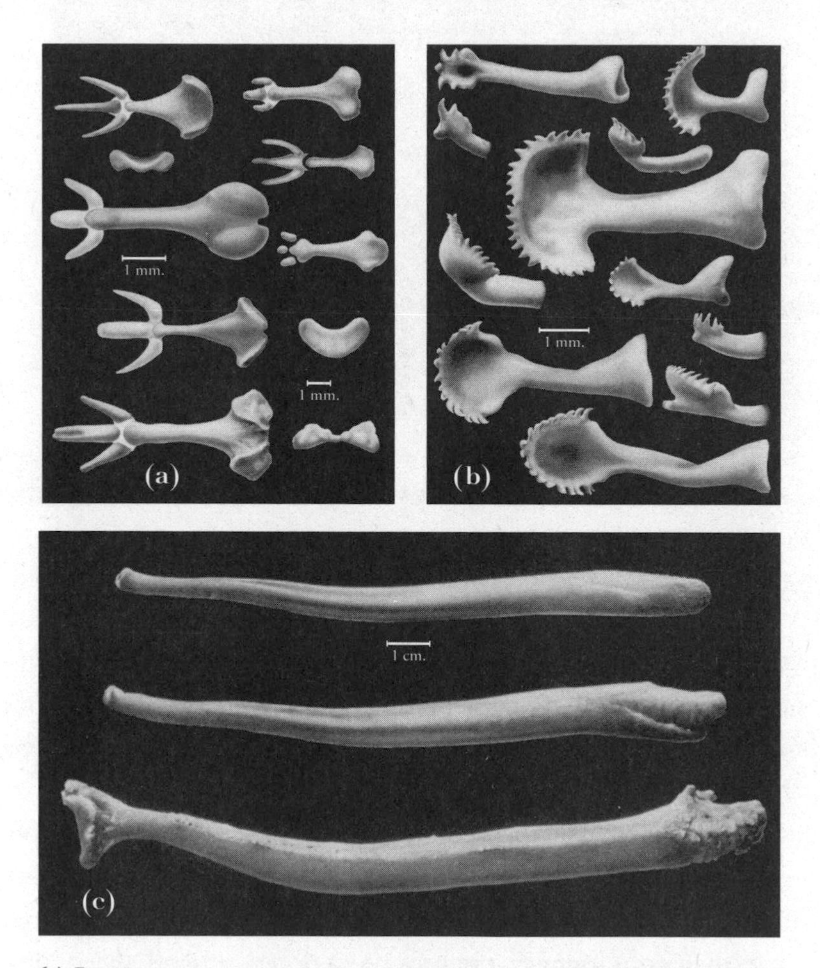


*(a) In voles, cotton rats, and muskrats, the penis bones are approximately one-quarter of an inch long. (b) In ground squirrels, they are slightly longer. (c) In bears and sea lions, they are 6 inches long, sometimes much longer.* —From William Henry Burt, "Bacula of North American Mammals." Miscellaneous Publications, no. 113, May 25, 1960. Illustrations by William L Brudon. Copyright © 1960 by Regents of the University of Michigan. Reprinted by permission.

*Chapter 2*

# BONE'S LIFE AND RELATIVES

An infant's shinbone is approximately 3 inches long. In adults, it is about six times that, depending on one's height. Throughout life, all bones maintain their unique shapes, but from fetal beginnings to the end of adolescence, they enlarge in all dimensions. How? Imagine a growing twig, which lengthens by continuously adding cells at its end. Bones, however, are capped with cartilage. If, like a twig, the bone grew longer by merely growing a thicker cartilage cap, a teenager's growing shinbone would be primarily cartilage. It would not be sufficiently strong and rigid to withstand the forces sustained by standing, much less skateboarding.

What actually happens when bone grows is nothing short of spectacular. Imagine spearing a small square of slippery banana peel (to represent freely gliding cartilage) onto the end of a growing twig. As the twig grows longer, the banana peel gets pushed ahead. The same concept applies to our bones. The cartilage caps remain relatively thin but get pushed ahead as growing bone fills in behind it. The extent of an individual's bone growth is partially the result of the person's unique genetic makeup—for example, tall children are usually the product of tall parents. If the genes go awry, bones may end up particularly short or particularly long, and an orthopedist may need to intervene. The extent of bone growth is also partially controlled by nutrition. Americans, in response to ample food, are

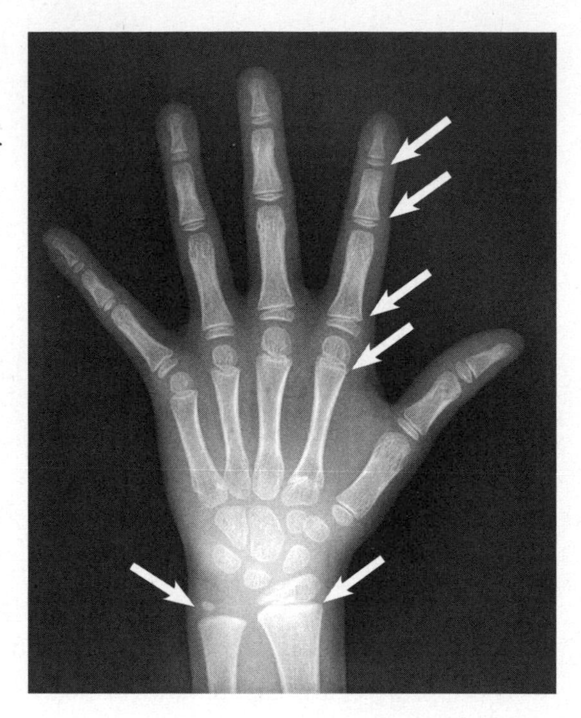

*Growth plates (arrows) are present at the ends of all the long bones in this 7-year-old's hand. From these areas the bones grow longer. In adolescence, as longitudinal growth comes to completion, each plate fuses with the bone's shaft.* BENJAMIN PLOTKIN, MD.

taller now than they were 200 years ago. The rate of bone growth is also influenced by hormones, which account for growth spurts experienced in puberty.

The area beneath the cartilage cap on the end of a bone is called the growth plate. Stimulated by hormones, it rapidly produces new bone cells during growth and pushes the cartilage cap ahead. The growth plate eventually burns out and disappears in late adolescence, typically sooner in girls than in boys. Growth plates on different bones fade away at different times during maturation, but always in a predictable sequence. Therefore, from an X-ray of a growing person's hand, orthopedists and radiologists can determine a growing person's age and time to skeletal maturity according to which growth plates remain active. (Also, anthropologists use the presence or absence of growth plates in various bones as a determinant of age at death.)

During periods of rapid growth, when growth plates are producing bone at warp speed, the new growth is particularly weak and vulnerable to fracture. These fractures almost always heal, but some injuries may irreversibly damage the growth plate. The consequences can vary and depend on the specific bone that has been injured and the age of its owner. If a 16-year-old martial artist fractured a finger bone through its growth plate and the bone stopped growing a sixteenth of an inch short of normal, the hand's function and appearance would be entirely normal. But if an 8-year-old skateboarder broke his thighbone just above his knee and the growth plate gave up the ghost, his leg could be as much as 4 inches shorter than its mate by the end of adolescence, producing major functional and self-image issues.

Repeated impact injuries, rather than all-of-a-sudden growth plate fractures, are also problematic. Consider preteen female gymnasts. When performing vault and floor exercises, these athletes run at top speed, leap, and land forcefully on their wrists before flipping and twisting back onto their feet. Over time, these repeated impacts can cause the major growth plate in each forearm to stop functioning. Because the other bone in each forearm continues growing, the ensuing length mismatch may cause wrist pain and perhaps even a career-ending deformity.

Our longest bones—those in the thigh, lower leg, upper arm, and forearm—have growth plates on both ends. But the growth plate on one end of each bone contributes more to that bone's ultimate length than does the one at the other end. The upper arm and lower leg bones grow mostly from the ends closest to the body's trunk, while the bones in the forearm and thigh lengthen predominantly from the ends farthest away from the trunk. Because these growth plates have greater or lesser effect on the ultimate length of the bone, orthopedists use a simple trick to remember which growth plates, if injured, are at particular risk for producing major limb-length discrepancies. Imagine sitting in a short, half-filled bathtub with your hands on your knees. The growth plates not submerged

*An imbalance of growth hormone during childhood and adolescence can lead to over- or understimulation of bone growth and have a profound effect on adult height.* WELLCOME IMAGES.

in the water—those at the shoulders, knees, and wrists—are the major contributors to longitudinal bone growth. Those submerged in water—the ones near the ankles and elbows—contribute less.

What about instances of dwarfism or giantism, when growth plates are universally affected? Mammalian bones are genetically predetermined to grow to a certain size and then stop. The controlling factor is pituitary growth hormone; too much of it, and activity in the growth plates rages. A pituitary gland tumor can send growth hormone levels soaring and lead to overgrowth. Such was the case for the iconic French wrestler and actor Andre the Giant. Conversely, an underproduction of growth hormone results in short stature. The limbs and trunk of a person so affected will be normally proportioned but just equally small in all respects. Tom Thumb comes to mind. Giving extra growth hormone to short-statured children is controversial, but it can boost growth plate activity and elevate stature into normal range.

The same genetically predetermined limit to bone growth affecting mammals also occurs in birds: all adult sparrows of a given species are the same size. However, this is not the case for fish, amphibians, and reptiles. Their growth plates never close, and their bones continue to grow throughout their lives, although more slowly after sexual maturity. By logic, this might lead you to believe that the huge python you spot in the tree overhead is a grandparent, but that is only partly correct. These animals' ultimate size is largely determined not only by age but also by the availability of food in its early life. With abundant prey, the python grew quickly during its early life, then continued to grow, but more slowly.

To withstand strain on the body—whether by jogging, snowboarding, or lifting heavy loads—bones must also grow thicker as they grow longer, and they do so without the aid of growth plates. In this instance, the twig analogy holds. Twigs thicken into branches as layers under the bark add new wood—an additional ring every growing season. Our bones thicken in the same way, except that they do not show obvious growth rings. Under normal circumstances, bone growth continues year-round and does not accelerate in summer months.

Still, bone does need sunshine—just not directly on it. Skin protects bone from such direct exposure and further aids bone by making vitamin D when exposed to sunlight. In turn, vitamin D facilitates the transfer of dietary calcium into the bloodstream. This system operates much like a bank—with money (calcium), officers (vitamin D), regulators (hormones), and its own Federal Reserve (parathyroid glands). Let me set the scene.

Recall that calcium is the principal element in hydroxyapatite, the crystal that makes bone hard. In a dissolved state, calcium is also essential to other tissues, including nerve and muscle. For instance, the heart, a highly specialized muscle, suffers if it is not getting precisely the right level of calcium. Too much, it gets twitchy. Too little, it gets cramps. Either one is life threatening, so the body has a finely tuned mechanism to keep the blood levels of calcium within the appropriate and narrow range. Glucose and carbon dioxide levels in our bloodstream can vary considerably, depending on when we last ate or how fast we breathe, but calcium levels normally never vary by more than 1 to 2 percent—akin to driving your car always at the same speed regardless of road conditions. The heart, that prima donna, insists on it.

So where does the calcium come from to keep the heart happy? If dietary intake is sufficient and steady, the needed calcium comes from what we eat and drink. But because none of us are likely to sip on calcium smoothies around the clock, our bones become a bank that loans calcium to the bloodstream and takes it back as conditions change. The bank's officers and regulators, vitamin D and several hormones, work together to keep the heart ticking happily, even if it means that the calcium reserves in our bones become depleted over time. When this happens, a banking crisis in the form of a fragility fracture can occur.

The main calcium-regulating hormone comes from the four parathyroid glands, located in the neck. Their power far exceeds their bulk. They are about the same weight, size, and consistency as soggy cornflakes and are pasted onto the thyroid gland, midway between chin and breastbone. Consider these quadruplets to be the Federal Reserve of Calcium, overseeing the calcium banking industry. The Fed constantly samples calcium levels in the blood. If an imbalance occurs, the parathyroid glands send hormonal orders to the gut, kidneys, and bone to delicately expand or contract the calcium level in the blood so that the heart can continue throbbing rhythmically and efficiently.

Although bone has an adversarial relationship with the heart, the more calcium the heart takes in from the gut and other sources means fewer bone loans must be made to satisfy the pump's calcium needs. Accordingly, awareness of the body's natural processes should encourage us to get outside more and trigger the replenishment of vitamin D. Vitamin production by sunbathing, however, is less effective in the winter and is thwarted by sunblock, advancing age, and pigmentation. For these reasons, dairies add vitamin D to milk, and your doctor may advise you to take a vitamin D supplement.

In adults, the calcium content of bone peaks around age 25 and then gradually diminishes, decade by decade. This waning accelerates dramatically in women after menopause. Weakened spine segments, which are originally short cylinders, can collapse into wedges, causing the back to bow, forming a dowager's hump. As conditions worsen, the inevitable diminution of balance, eyesight, and agility may lead to dangerous stumbles and falls. Broken hips and wrists in fragile bones abound—but still, the heart beats merrily away, blissfully unaware of the structural damage its hefty demands are causing. Even worse, the heart is oblivious to the fact that its safety depends on the rigidity of the bony cage that surrounds and protects it. Without sturdy ribs, spine, and breastbone, every hug would be a brutal squeeze on that prima donna of a pump inside the chest.

In children who have calcium or vitamin D nutritional deficiencies, the heavy demand on the bone bank results in rickets, which is manifested by soft bones, painful joints, and legs that become bowed. Rickets was rampant in the northern industrial slums of America and Europe around the turn of the twentieth century. The smoke from coal fires used for heating and for powering industry blocked the sun's ultraviolet rays, which became further blocked by tall tenement buildings springing up along crowded cities' narrow streets. Investigators eventually learned that the disease's skeletal devastation could be prevented by eating butter, animal fat, or cod liver oil. In 1922, a nutritional investigator at Johns Hopkins Uni-

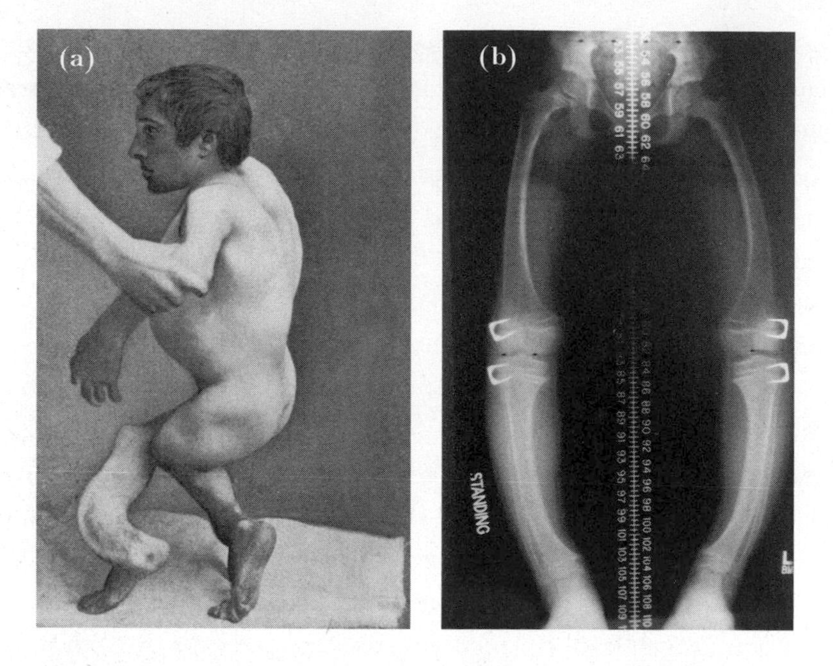


*(a) This 1912 photograph shows the effect of rickets on bone strength, particularly evident in weight-bearing bones. (b) In modern times, staples placed laterally across the growth plates retard the progression of bowlegs. Growth on the unstapled sides of the knees will correct the bowing over time.* (A) WELLCOME IMAGES; (B) MUSEUM OF MAN.

versity identified and named the active compound in these "foods" that could prevent and treat rickets—vitamin D.

✦✦✦✦

IT IS TIME to examine how bones remodel themselves, something that hearts, those mere pumps, cannot do. If the heart survives an attack, the damaged area haphazardly heals with a scar, which may impair function in the surviving cardiac muscle. Contrast that type of healing to what occurs in bone. When attacked and strained or even cracked, bone heals itself completely and without scarring.

Earlier I mentioned that a bridge, once built, cannot increase its carrying capacity. Bone can. How does it achieve such a miraculous and most convenient feat? I previously introduced the osteoblasts, which make bone. Their foils are the osteoclasts. *Osteo* of course means "bone," while *clast* means "breaking down." In short, osteoclasts destroy bone. In our banking analogy, they are thieves, who work overtime when their ringleader, the heart, needs a calcium fix. Osteoclasts also go to work every time stress is applied to a bone.

Let's see how this amazing process works. Within our bones osteoclasts and osteoblasts organize themselves into assemblages known as cutting cones. Each one is similar conceptually to a giant boring machine that digs subway tunnels under cities and is followed by materials and equipment that line the new tunnel with concrete. The lining both supports and seals the tunnel walls. Later, if cracks occur, the tunnel can be relined repeatedly, at least until the passage becomes too narrow for the train to zoom through. You have millions of microscopic versions of these tunnel makers in your bones, each working to keep them strong and responsive to the mechanical stresses that we apply to our skeletons. The cutting cones are tipped with bone-dissolving osteoclasts, which are constantly boring minuscule tunnels through our bones. Trailing behind the osteoclasts, osteoblasts line the walls of the tunnel with concentric rings of new bone containing crisscrossed layers of collagen, which predates the concept of plywood by many millions of years.

When the bone tunnel lamination is complete, only a tiny central channel remains. During life, these channels contain small blood vessels that nourish the cells trapped between the laminated layers. These tunnels were first discovered and identified in 1691 by British physician Clopton Havers, who observed them with a magnifying lens and described them in his protractedly titled book *Osteologica nova, or Some New Observations of the Bones, and the Parts Belonging to Them, with the Manner of Their Accretion and Nutrition*. We appropriately call these tunnels Haversian canals.

The cutting cones not only produce new laminated bone but are also constantly at work refreshing and remodeling older bone.

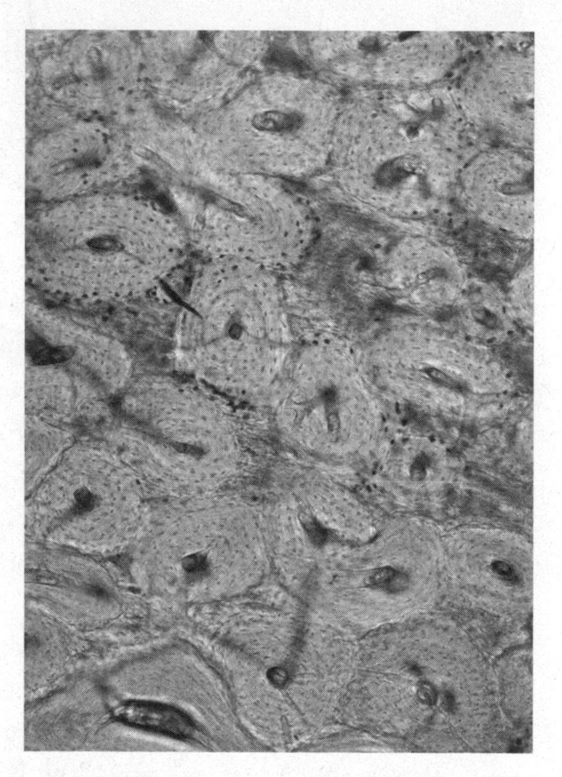

*Under the microscope, compact bone consists of multiple cylinders, which are seen here end on. Each one is roughly three times the diameter of a human hair and made from concentric layers of bone surrounding a central tube. Blood vessels pass through each tube to supply nutrients to the encased bone cells (small black dots).* KEVIN MACKENZIE, UNIVERSITY OF ABERDEEN; WELLCOME IMAGES.

Imagine dropping a red-hot stone onto the surface of a frozen pond. It melts a hole through the ice and sinks. The water filling the hole then freezes to make a plug of new ice. Keep scattering hot stones one by one across the pond, and eventually all the ice will be replaced. In the process, some of the newly formed ice plugs may be partially or completely replaced several times over, depending on where each stone lands. On a microscopic scale, this remodeling process is occurring continuously in our bones, courtesy of the cutting cones. Bore a hole, fill it in. Bore more holes, fill them in. Hot rocks tossed on the ice land randomly, but cutting cones don't. They go where the bone needs strengthening. How do they get their directions?

So far, I have tried to share my passion for bone without resorting to jargon. Unfortunately, I do not know how to get around introducing the term *piezoelectric*. *Piezo* in Greek means "squeeze" or "press" (to help remember both the approximate pronunciation and the meaning of piezoelectric, think *squeeze-o-electric*). When certain crystals are compressed, they develop an electric charge. This is the piezoelectric property, which hydroxyapatite possesses. When bone is compressed, the hydroxyapatite crystals become faintly electrified. In fact, every time you take a step, the portions of your bones that are resisting gravity are producing piezoelectricity. The cutting cones sense these electrical fields and head toward them. Consider, for example, what happens during a game of tennis. The bones in your racket-swinging arm experience a jolt every time you hit the ball. Each stroke generates piezoelectric charges, which alert the cutting cones to start working in the "quake zone" to make new bone to resist these unaccustomed forces. Of course, there will be far fewer piezoelectric charges firing in your opposite arm, so the cutting cones over there remain on standby. Slowly the cutting cones will sufficiently remodel the bones in your racket arm such that those bones are measurably denser and thicker than the ones in your opposite arm. A perceptive German surgeon named Julius Wolff made this observation in the late 1800s, before such changes could be seen on yet-to-be-discovered X-rays. The phenomenon, known now as Wolff's law, is simple: bone responds to applied stresses. Within the cutting cones' capacity, more stress leads to strength; less stress leads to weakness. Tennis, anyone?

The very reason we are encouraged to exercise to maintain bone mass is that physical activity generates barrages of piezoelectric forces. Even a gentle stroll will stimulate piezoelectric charges in the lower limbs, pelvis, and spine. The cutting cones sense these electrical messages, recognize the need for the bones to resist the mechanical forces of walking, and strengthen the bones that are experiencing the repetitive loading. Stand up and bounce on one foot for a moment. Appreciate that your cutting cones sense the message and will respond by making strong bone. Hydroxyapatite needs a bit of a jolt to generate a piezoelectric force, which is easily triggered by moderate-impact activities such as jogging or brisk walking. While swimming and cycling are healthful in many ways, they don't jolt the bone sufficiently to stimulate the cutting cones.

For those who are unwilling or unable to stimulate critical piezoelectric forces by walking, would it work to stand in the back of a pickup truck while it rattles down a gravel road? That might work, but gravel roads and pickup trucks are not always available. Tamer options exist, including vibrating platforms that can be bought online with two clicks. The manufacturers of these stand-on devices make all sorts of claims regarding the purported health benefits from regular use, including increased bone density. While some scientific studies support such claims, the different platforms vary widely regarding the vibrational rate, intensity, and direction as well as the recommended frequency and duration of treatment. Several studies have failed to find that the platforms affect bone density. I am sticking with walking, but the vibrating platforms might be helpful in preserving bone mass in individuals who are unable to walk and resist gravity normally—children with severe cerebral palsy and astronauts, for example.

Just as bones respond microscopically to applied stresses, growing bones change shape in response to bending forces. When children first begin to walk, they are often noticeably bowlegged and swagger like little cowboys. The alignment normally corrects itself over the next several years as the growth plates around the knees produce slightly more length on the inner sides of the knees than on the outsides. This differential growth straightens out the legs and even slightly overcorrects the bow. Most adults are therefore slightly knock-kneed. Toward the end of growth, if bowlegs or excessive knock-kneeing persists, an orthopedic surgeon can operate to retard or stop further growth on the side of the bone that is too long, giving the short side a chance to catch up and correct the angulation.

Infants and toddlers on every inhabited continent have had their skulls intentionally shaped starting as far back as 45,000 years ago. Anthropologists can only speculate why various cultures performed this ritual. It may have been to denote elite status or ethnic identity, but the reasons may be as diverse as the cultures that practiced it. Some Native American tribes made the back of the skull flat by

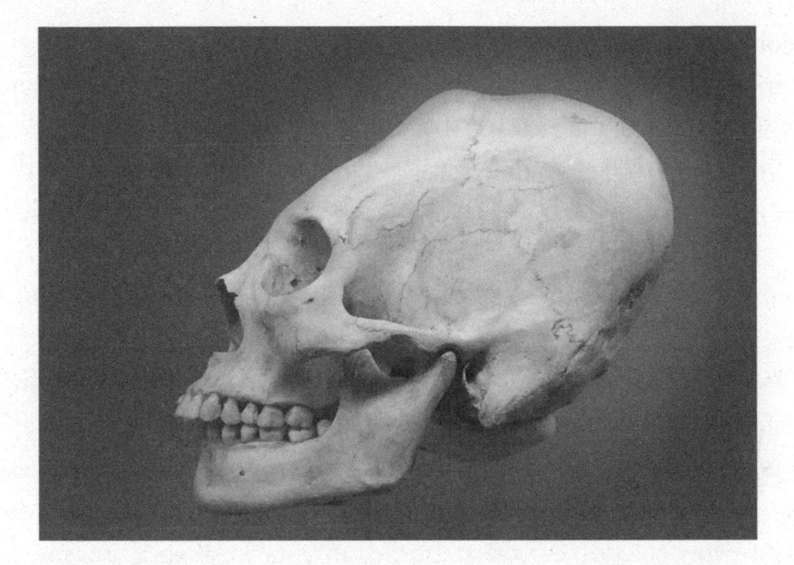

*This skull from Peru is over 2,000 years old. Cultures on every inhabited continent have practiced skull shaping for reasons that are presently obscure. The custom demonstrates the plasticity of growing bone.* MUSEUM OF OSTEOLOGY.

binding an infant's head securely to a cradleboard. The infant's skull bones and the joints between them were pliable, so the skull slowly flattened against the board. By age 3, the shape was permanent. Other cultures, including the Huns, Maya, and Pacific Islanders, preferred the look of an elongated skull. They achieved this shape by binding a cord around the infant's head from just above the eyes in front to the nape of the neck in back. This pressure gradually molded the soft bones and the intervening pliable joints into the desired shape.

Contrast this process of skull shaping to the long-practiced Chinese custom of foot-binding, performed on girls between 4 and 9 years of age. A combination of broken bones and progressively distorted joints was required to achieve the favored appearance.

Personally, I cannot say I'd enjoy the process of foot-binding or skull shaping, but they speak well for the versatility and adaptability of bone, which has to compete for the title of the world's best building material with some other amazing substances. Consider those hard things that protrude through the gums on most vertebrates and through the feet on hooved animals. Are they bone? In the instance of teeth, no. Although teeth and bone are both hard and calcium dense, they have entirely different chemistries and structures. So don't add teeth to the total-body bone count of around 206. The same goes for ivory tusks, which are continuously growing front teeth.

What about hooves, claws, and fingernails? Yes, all are potentially dangerous weapons and definitely have protective value, but they are not bone. Rather, they are all made of keratin, which is another fibrous protein with some similarities to collagen. Since keratin does not have any calcium crystals deposited into its meshwork, it is more flexible and lighter than bone. For instance, we have a fine meshwork of keratin in our skin, which makes it a tough protector of our precious bones. Thicker layers of keratin cover the bone in turtle shells, bird beaks, and cow horns.

A point of confusion surrounds the term *whalebone*. It can refer to a whale's actual bone, or it can refer to baleen, which is present in some species of whales that use it to filter tasty krill from big gulps of seawater. Baleen is made out of keratin, and in the nineteenth century, long flexible strands of it were made into collar and corset stays, buggy whips, and umbrella ribs. Today, strips of steel, plastic, and reed rather than baleen reinforce and give shape to corsets, hoopskirts, and costumes, but the strips and the requisite sewing are still called boning.

Although keratin is exposed and visible, living bone typically is not. There are two exceptions. The most obvious one is antlers. Male moose, reindeer, and other members of the deer family grow and then shed antlers, usually annually. They sprout from the skull and grow faster than any other mammalian bone. As usual, an antler grows longer from the area immediately beneath the tips' cartilage caps. During growth, a thin layer of velvety skin, rich with blood vessels, covers the antler and supplies the needed nutri-

*Right whales (pictured) and humpback whales use baleen to filter food after gulping large volumes of water. Although it is sometimes called "whalebone," baleen is made of keratin, which also makes up hair and fingernails. Inuits used the jawbones of right whales to frame their living quarters.* French National Museum of Natural History.

ents for rapid growth. Once growth is complete, the animal rubs the velvet off; and deprived of blood and nourishment, the bone dies. Then the bare-bone antlers are out there for all to admire until osteoclasts, the bone-destroying cells, release their attachment to the skull. Left on the ground, the shed antlers are an excellent source of calcium for Earth's little creatures, who gnaw on them to keep their hearts happy.

The other exception to naturally exposed bone is the shield-like skin on some reptiles and frogs and on a few mammals. This specialized skin is called osteoderm (combining the words *bone* and *skin*). It provides a flexible defensive armor and accounts for a stegosaurus's huge spinal plates, a Gila monster's bumpy skin, and for the impressive shield on an armadillo. Crocodiles make two additional uses of osteoderm. When the croc is out of the water on a cold day, its bone skin, which has a great blood supply, is a solar collector and warms the owner's innards. On a hot day, the heat reverses its course through the osteoderm. Or the beast may decide to totally submerge and hold its breath. The osteoderm retards carbon dioxide accumulation and blood acidification by temporarily exchanging bad ions in the blood for good ones in its store.

I mentioned osteoderm for the sake of completeness. It provides protection, and for crocs thermoregulation, but osteoderm does not offer any skeletal support. Accordingly, let's return to support.

Have you ever seen a 2-inch-diameter worm? Of course not, because they are nonexistent. Worms and snakes are both generally cylindrical, and there are certainly 2-inch-diameter snakes. The question thereby arises, Why can snakes get big and scary but worms can't? Not that I am advising it, but if you stepped on a fat worm, you might hear a squish. If you stepped on a fat snake, you might get bitten while listening to the crackle of myriad ribs breaking. The snake's bony skeleton supports its weight, which allows it to move quickly. In times gone by, bone even supported dinosaurs nearly half as long as a football field and four stories high. A worm, by contrast, does not have any rigid internal framework. It cannot resist gravity in a major way and therefore remains lowly. Sure, sea jellies and other boneless water-bound animals can get big, but that is because the ocean supports them against gravity's crush. There are several other successful skeletal support strategies, not for dinosaur big, but certainly worth appreciating.

Snails, clams, and oysters find rigid support from an outer shell of calcium carbonate, one form of an exoskeleton. Each owner keeps adding calcium carbonate to the edge of its shell so it can keep on growing while remaining continuously protected and supported. The shell, however, is heavy. Giant clams have tipped the scales at 600 pounds, nearly all of it calcium carbonate.

Coral polyps remain the same size (about as big as a grain of rice) throughout their lives, and they secrete calcium carbonate for their external skeleton just as snails do. Individual coral organisms pool their efforts and form a communal reef—protective but not mobile. Nor can they chase prey or escape prey. Can other animals do better?

Insects and their cousins—spiders, crabs, shrimp, and lobsters—have external skeletons of a different type. Their crunchy exoskeletons are made of chitin, which consists of long chains of molecules derived from glucose. A chitin exoskeleton offers its owner shield-like protection. It is more waterproof than skin, which may be an advantage, especially for life out of water. The chitin exoskeleton arrangement works for tiny animals, including mosquitoes, gnats, and mites. By contrast, the smallest vertebrate, meaning an animal with a bony internal skeleton, is a rain forest frog measuring just over a quarter of an inch from nose to rump. If the goal is to remain small but protected, chitin rules.

Chitin skeletons are also lightweight, so chitin-clad creatures can move faster than snails, and some even fly. The design is clearly successful, since there are at least 10 times more insect species on Earth than all other animal species combined. But a chitin skeleton has two problems. The first is that it cannot enlarge, so periodically its owner has to squirm out and remain vulnerable to predators, including those of us who enjoy a soft-shell crab sandwich, until it can secrete a new, larger covering. The second problem is that chitin can support only a limited amount of body weight, especially on land. You may be able to find a 4-pound lobster, but fortunately there are no 4-pound spiders walking around. And except for a 2-foot-long walking stick insect, you could hold the world's largest insect in your palm (lucky you). Contrast that to the largest terrestrial dinosaur, estimated to have weighed 50 tons, the equivalent of eight or nine elephants. Therefore, it seems that if you want to be big and terrestrial, not only do your innards need some rigid support, but that support needs to be internal and made of, what else, bone.

I must digress for a moment to be biologically correct. Vertebrate animals are defined by the presence of a spinal cord and include fish, amphibians, reptiles, birds, and mammals. In most instances, the spinal-cord-encompassing spine is made of bone. The exceptions are sharks, skates, and rays. Rather than bone, their skeletons are made of cartilage, which is lighter and far more flexible than bone, like the rubbery consistency of your nose and outer ear. These properties of cartilage make cartilaginous fish highly efficient swimmers. Sharks are also highly efficient eaters because calcium deposits harden their jaw cartilage to produce a powerful bite. This calcified cartilage is not bone but might seem so at first glance. This should be reassuring in case one of these awesome creatures swims near.

For the most part, however, vertebrates have skeletons made from bone. Not only can some vertebrates grow beastly in size, but since the bones are hollow and therefore relatively light, they also allow their owners to move quickly. Both endoskeletons and exoskeletons have their success stories. I am fine with my endoskeleton, and in no way am I envious of a snail's or bug's means of support and protection. I don't have to shed my skeleton periodically, and unless I get stomped on really hard, my skeleton does not crack. Yet nothing is perfect. Even bone can fail.

*Chapter 3*

# WHEN BONES BREAK

WHAT HAPPENS IF YOU BECOME ZEALOUS IN your endeavor to increase your bone density by hiking 10 miles every day with a 40-pound pack? You may well get a fatigue fracture, also known as a stress fracture (or in the case of military recruits, a march fracture). This is what happens.

Bend a coat hanger wire once, and nothing happens. Bend it repeatedly in one spot, and first the paint peels off, then the shiny metal goes dull, and finally the wire gets hot and breaks. The same thing can happen to bone. Cutting cones can only work so fast. If you repeatedly stress the bone faster than the cutting cones can reinforce it, the bone gradually weakens. Microscopic cracks develop and cause local pain and tenderness, particularly during the offending activity. Initially, the crack is so small that the X-rays remain normal, although magnetic resonance imaging (MRI) would show extra fluid in the injured area.

Fatigue fractures can occur in army recruits' feet, runners' shins, dancers' hips, and gymnasts' wrists. Avoiding the offending activity for a few weeks will usually let the cutting cones gain and then maintain a lead if the activity intensity is thereafter increased gradually. Conversely, ignoring the pain and continuing to stress the bone can cause it to fail completely—snap, just like the coat hanger. Without employing welding skills, the wire remains broken. The amazing bone, however, has the capacity to heal itself, likely without any trace of the injury left behind.

So how does a complete fracture heal? Let me start by clarifying some terminology. To orthopedists, *fracture* and *break* mean the same thing. I often hear, "What? You just said my bone is broken. The emergency room doctor said it was fractured," or vice versa. I am not sure whether some people think that a fracture or a break is worse or where the supposed differentiation originated, but consider the words synonymous.

Now that I have cleared that up, let's say you stumble over an unseen parking barrier and instinctively throw your hands out in front of you to avoid bashing your face on the pavement. As you sit up, you notice big pain in your wrist and that your hand is pointing northwest off the end of your northbound forearm. The bone snapped and the fragments are angulated, a major seismic event in the life of a bone. What would happen if you went home, wrapped a magazine around your deformed wrist, and secured it with a necktie? No emergency room or orthopedist for you. Instead you'll just let nature run its course. This scenario seems implausible these days, at least in industrialized countries, but animals have been healing fractures without professional attention for millions of years. There are fossils and penis bones to prove it. What happens if your broken wrist is left to heal unattended?

Those wonderful cutting cones would like to get busy just like they do for fatigue fractures, but this time there is a huge difference. The gap between the jumbled, jagged fragments is too big for the cutting cones to cross—in the same way that a desert tortoise, rattled by a trembler, cannot leap across the San Andreas Fault. If we temporarily filled the rift, however, with some semisolid substance, say mud, the tortoise could amble to the other side, even on shifted ground.

What actually happens with broken bones continues to astound me. While the cutting cones bide their time, blood from torn capillaries leaks out immediately and fills the fracture gap. Then over the span of 2 weeks, new capillaries and a collagen meshwork form

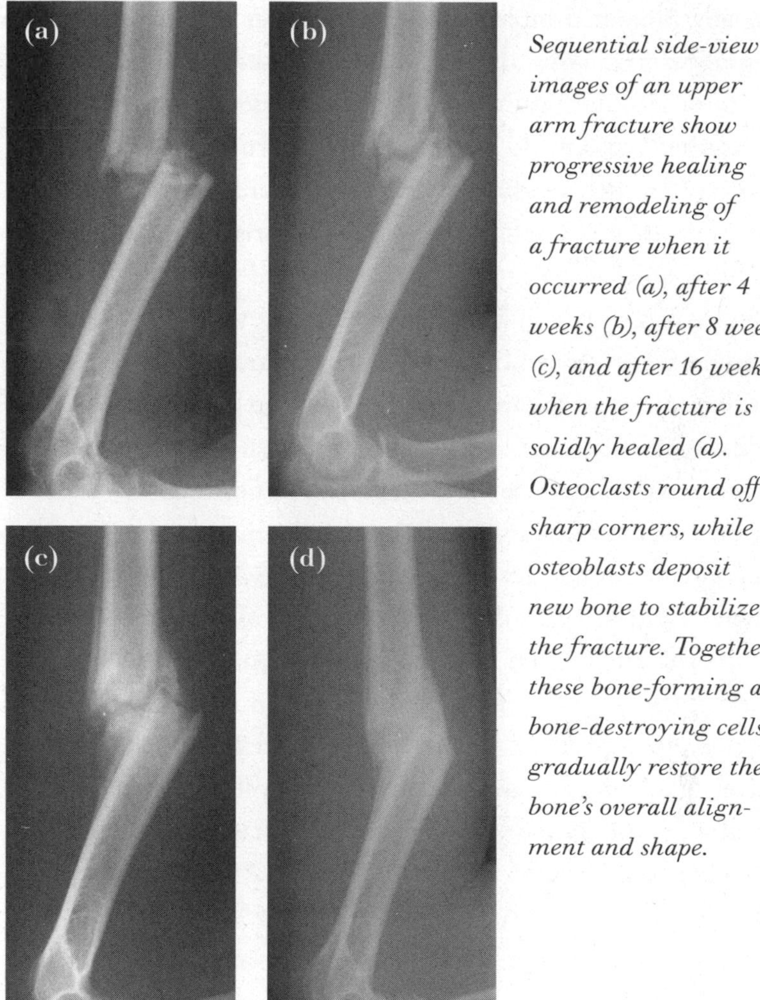


*Sequential side-view images of an upper arm fracture show progressive healing and remodeling of a fracture when it occurred (a), after 4 weeks (b), after 8 weeks (c), and after 16 weeks, when the fracture is solidly healed (d). Osteoclasts round off sharp corners, while osteoblasts deposit new bone to stabilize the fracture. Together, these bone-forming and bone-destroying cells gradually restore the bone's overall alignment and shape.*

in the clot. Also, soon after injury, various chemical alarm sirens begin sounding. The cacophony awakens nearby cells that begin to produce cartilage, which is puttylike in consistency. Osteoblasts in the area contribute a bit of primitive bone to harden the putty. After 3 to 6 weeks, depending on the size of the bone and the size of the gap, the shaking is over. Voilà, the fragments are tentatively stuck together with new bone.

Now the cutting cones go into action. They follow the piezoelectric signals generated in the primitive bone and lay down strong, mature bone by drilling thousands of holes and then filling them in. For most fractures in the hand, bone strength has recovered sufficiently after 4 to 6 weeks to resume sports and manual labor. For lower limb fractures, because healing has to resist weight-bearing, it may be 12 to 16 weeks before you want to challenge another parking barrier. The cutting cones continue to remodel the bone for many months, even for years in large bones that had major displacements of the fracture fragments. Evidence of the original damage gradually diminishes and may even disappear as the cutting cones complete the reconstruction.

Such is not the case with most other tissues. For instance, decades later, you can probably still see your childhood skin scars. And if the heart suffers an attack, the damaged muscle turns entirely to scar, which will forever affect the remaining muscle fibers' ability to pump. The only tissues that can heal without scar are bone and cornea, but the story of cornea will have to wait.

Patients often ask their doctor, "What can I do to help my fracture heal quickly?" The short answer is, keep the osteoblasts and osteoclasts happy. They are working around the clock at the fracture site to make and remodel new bone. If they are not receiving the building materials they need, construction lags.

There are three types of factors that affect fracture healing. Ones that are at least partially under the patient's control include nutrition (eat a well-balanced, healthy diet), smoking (don't), diabetes (control it), and infection (avoid it, or if present, treat it aggressively). Malnutrition adversely affects all types of wound healing because the entire body is deprived of necessary nutrients, and there are certainly not extra ones for shipment to the construction site. Nicotine constricts blood vessels, so even if there are adequate nutrients in the region, the narrowed blood vessels cannot properly deliver the goods. Diabetes also diminishes blood vessels' ability to supply nutrients to the construction site. Additionally, wildly fluctuating blood sugar levels from poorly controlled diabetes first jack up the repair cells with a

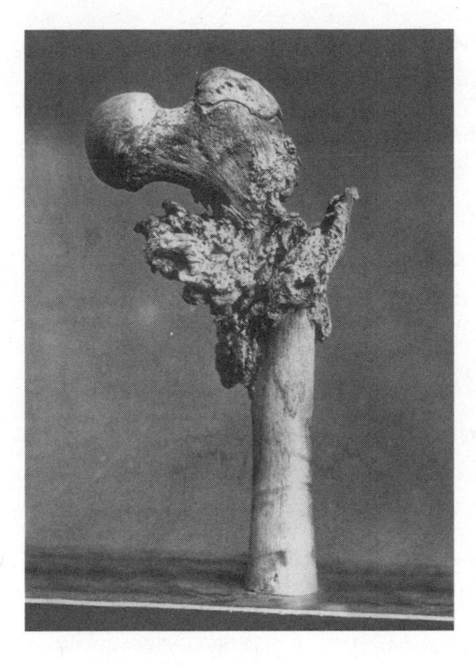

*At Gettysburg in 1863, a musket ball shattered this thighbone in an 18-year-old soldier. Despite an uncontrolled infection that caused his death 10 weeks later, the injury produced exuberant new bone in an effort to heal the fracture.* Otis Historical Archives, National Museum of Health and Medicine, photo ID NCP 1603.

sugar high and then snatch the treats away. The cells get confused. Then if an infection occurs, bacteria compete with the osteoblasts and osteoclasts for available nutrients. Infected fractures can heal if managed properly, and a patient can minimize an infection's damaging effects by complying with the doctor's directions regarding wound care and antibiotic treatment.

The second category of factors affecting healing are ones that are at least partially under doctors' control. These include motion at the fracture site, diabetes, and hardened arteries, which restrict blood flow. If a fragile capillary, which is trying to grow into the fracture site and deliver nutrients, gets ripped asunder by movement between the bone ends, then nutrients are unavailable. With medications, doctors have some influence on controlling diabetes, dilating creaky blood vessels, and improving the heart's pumping efficiency, all in an effort to facilitate nutrient delivery. Steroids are potent anti-inflammatory medications, and inflammation is a necessary part of fracture healing. Therefore, steroids should be avoided when possible in order to promote rapid healing. Yes, medications such as ibuprofen

(Advil, Motrin), naproxen (Aleve), celecoxib (Celebrex), and aspirin are also anti-inflammatory drugs, although not nearly as powerful as steroids. In certain laboratory experiments, these nonsteroidal anti-inflammatory medications do adversely affect wound healing, but their effects in real-world fracture healing appear to be negligible. For that reason, modest use of nonsteroidal anti-inflammatory drugs for pain control while convalescing from a broken bone is better than scarfing narcotic pills.

The third factor that retards bone healing and that is not under anybody's control is advanced age. Just as old folks are not as spry as youngsters, neither are their osteoblasts and osteoclasts.

Fractures that are slow to heal for any of the mentioned influences may benefit from attending "a rock concert for bones." Just as a loud clap of thunder can rattle teacups and just as a rock concert woofer can shake your innards, sound waves can also make your osteoblasts tremble. The frequency of these sound waves, however, is too high for us to hear; it's ultrasound. A person with a fracture can place the ultrasound unit (about the size of a deck of cards) on the skin over the fracture and treat the local osteoblasts to a rock concert that only they can sense. Used 10 to 20 minutes a day, this barrage of sound vibration creates piezoelectric forces in hydroxyapatite similar to what occurs with walking. In response to the rattling and shaking, bone-forming cells step up their work tempo from waltz to rock and roll—a biological response to mechanical stimulation. To prove the effectiveness of ultrasound on fracture healing and to avoid any preconceived notions affecting the results, investigators conducted "double-blind" studies, meaning that neither the researchers nor the patients knew until the studies were over which subjects had received the real deal and which ones had received sham treatment. Those patients receiving the ultrasound treatment showed significantly faster fracture healing than those receiving the "sugar pill." So even though we cannot see, hear, or feel the ultrasound, the healing bone knows and enjoys the vibes.

Would the bone respond similarly to electrical or magnetic stimulation? There is some theoretical plausibility for their use, since we know that compressing hydroxyapatite generates piezoelectric forces and that fluctuating magnetic fields also create electric currents. Investigators, entrepreneurs, and charlatans (not necessarily exclusive of one another) have tried using magnetic fields and various sources of electrical stimulation to stimulate bone healing. In the distant past, enterprising folk used electric eels and lodestones for such purposes. The more recently available, typically patented devices have either surface units or electrodes implanted into the bone near the fracture site. The promoters of these devices, however, have shunned testing the effectiveness of their devices in double-blind studies. Without proven effectiveness, interest in these devices has waned. Admittedly, the concept is more acceptable than trying to keep a squirming eel up one's sleeve.

What happens when a large section of bone is missing after a severe fracture or removal of a tumor? The body tries to fill the gap with new bone as it does for routine fractures, but the void and movement may be insurmountable despite how hard the bone tries to repair itself. The gap instead fills with gristle-like fibrous tissue, the bone ends remain unstable, and a false joint forms. Orthopedists can span the gap with new bone brought in from afar. To help you understand bone grafts, consider the analogy of borrowing money.

If you find yourself just a couple of dollars short, searching under the sofa cushions and raiding the piggy bank may produce the necessary cash. Nobody misses the money, and there is no need to pay off the loan. If you need a substantial amount of money, perhaps you can borrow from your retirement account or from your kid's educational fund. Doing so solves the immediate financial crisis but leaves a deficit elsewhere, which may or may not recover over time. Finally, if you cannot fund it yourself, you could go to the bank and ask for a gift—yes, a gift.

The concepts are the same for bone grafting. If surgeons only need some robust cells to stimulate new bone formation, maybe to supplement local bone during a spinal fusion procedure, for example, then they can temporarily open the hard outer surface of the pelvis and scrape spongy bone from the pelvis's interior. Several tablespoons of bone are available, and its absence does not change the pelvis's shape. The crumbly graft so obtained offers no mechanical stability, but it is full of bone-forming osteoblasts that quickly overcome the small debit in the recipient bone. At the same time, the donor site fills with new bone and can contribute again later if necessary.

Sometimes a surgeon needs a short section of structurally sound bone to fill a gap in a nonhealing fracture site or to span a void after removing a bone tumor. In these instances, a chunk of full-thickness bone from the patient's own pelvic rim works well. Unless the owner is quite thin or the chunk is bigger than an inch square, the withdrawal is harmless and invisible.

If a long, straight bone graft is needed, attention often turns to the leg. Of the two bones between the knee and the ankle, the sturdy one that transmits weight is the robust shinbone, aka tibia. Just to its outside is the fibula, half an inch in diameter. Except for a short portion near the ankle, the fibula is merely an attachment site for ankle and toe muscles, which remain fully functional even in the fibula's absence. Hence, the fibula is a mainstay in the orthopedist's "long-bone lumberyard." Depending on the patient's height, a 6-to-10-inch strut of fibula can be used to span a large gap in a critical bone. The fibula strut is far skinnier than most of the bone segments it replaces, so it will need support from a sturdy internal plate and an external brace for at least a year. This bridging graft is "seed money" and will grow stronger to fully resist bending, twisting, and compression forces, a process that is not complete for several years.

Orthopedists have several ways to coax the fibula into earlier service. If the gap to be bridged is less than half the fibula's length, the fibula strut can be cut in half and doubled. Although it will still take many months for the segments to thicken enough to provide adequate support, it is less time than with using a single piece. In other words, it doubles the loan's productivity without increasing the debt.

Whatever its size, getting a loaner part into service quickly also improves its appeal. An orthopedic surgeon can kickstart a fibula graft into service by meticulously harvesting the bone along with the blood vessels that supply it. After the bone graft is secured in the gap, the surgeon connects the fibula's artery and vein to nearby vessels. Blood then flows through the fibula just as it did before the bone was moved. With the immediate restoration of circulation, the grafted fibula heals and enlarges far more quickly to pay off the loan.

The bone loans described so far all come from the patient's own body, so the patient's immune system raises no concern and there is no risk of rejection. At times, however, huge segments of bone are required and constitute loans too large for patients to make. Consequently, the surgeon may turn to a gift from a recently deceased organ donor. The bones are taken following removal of the heart, liver, and kidneys, which require placement on ice, immediate transplantation into a grateful recipient, and lifelong protection against immune rejection with powerful, rather risky antirejection drugs. By contrast, the bones undergo a leisurely cleaning to remove all the blood and protein they contain. Then they are dried, sealed in plastic bags, sterilized, cataloged, and shelved until needed. Absent their proteins, bones generate no immune response when implanted into a different person, so they can provide grafts of any size and shape without any risk of rejection. This is a wonderful gift, but it comes with some strings attached. Since the grafted cadaver bone has no blood supply and no cells, the recipient site has to supply these, which it does, but ever so slowly. In the meantime, the cadaver bone graft can crack, crumble, or dissolve. For that reason, grafting cadaver bone is not undertaken lightly.

If you happen to have an identical twin, you could borrow cell-rich living bone and do so without risk of rejection, since you both have the same immune system. Careful, though—your twin might need one of your kidneys sometime.

Just as you weigh the options before borrowing money, orthopedists discuss with patients the pros and cons of one type of bone graft over another and balance the relative merits of each. Spongy bone is readily available without causing any permanent skeletal defect, and it heals faster than compact bone; but compact bone is immediately stronger than spongy bone. The surgery is much longer and more difficult when moving bone and immediately restoring its blood supply, but healing time can be markedly shorter. Cadaver bone comes in any desired size and shape but is slow to heal. Sometimes the trade-offs are reduced by using two types of bone graft for the same debt—taking out a home loan *and* accepting a unique gift.

Mending damage from injuries constitutes a considerable part of a general orthopedist's practice. The goal is to facilitate a fracture to heal properly and quickly. In times gone by, healing in most any alignment was by itself considered a victory, and villainous foes—smoking, malnutrition, diabetes, and advanced age—have always been and will likely always be bugaboos. New methods of fracture fixation and osteoblast stimulation, however, continue to offer better results even in face of such fearsome adversaries.

These days, fracture healing with inadequate realignment of the fragments is a false victory. For instance, a child (and her parents) would not tolerate a healed fracture where the thighbone was straight but rotated such that her foot pointed sideways. Some less obvious fracture misalignments, however, may take years or even decades to reveal themselves. Consider, for example, a fracture through a joint where the contact surface is no longer smooth all the way across. Think of the problems created by highway pavement where one lane is 2 inches higher than the other. Cars can still pass, but something bad will happen eventually. If a fracture heals with such a step-off on the joint surface, the irregularity will gradually destroy the cartilage. Then the unyielding, underlying bones begin rubbing on each other, producing pain and swelling, limited motion, and maybe even squeaks and crackles.

Not every fracture presents such a grim picture. Indeed, sometimes what looks to the untrained eye like inadequately aligned

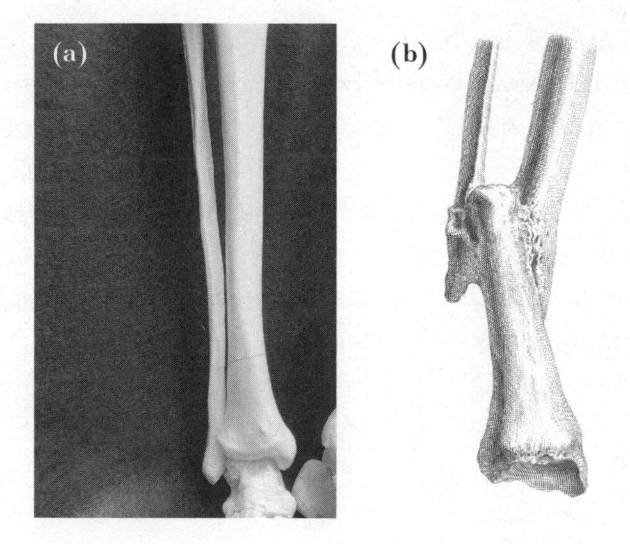


*The alignment and shape of normal leg bones (a) contrast sharply with broken ones (b), where it was once considered a success if the fracture healed, even if the leg was short, deformed, and caused a limp.* (B) William Cheselden, *Osteographia, or the Anatomy of the Bones* (London: W. Bowyer, 1733).

fragments provides the ideal outcome. Take, for instance, a fracture through the middle of the thighbone in a child. His healing response is so robust that the increased circulation working at the fracture site also stimulates the growth plates at the ends of the bone. If such a fracture is realigned perfectly, it heals, the X-rays look hunky dory, and the parents are happy—for the moment. Over the next year, however, the thighbone may outgrow its mate by up to an inch because of the extra nutrition and circulation in the thigh that the fracture generated. At maturity, the boy or his parents would not likely notice a half-inch difference in leg length, and he would be fully functional. (Even uninjured legs are often not exactly the same length, and nobody notices or cares.) An inch or more difference in leg length, however, causes the owner to list to one side, which he unconsciously corrects by bending his spine away from the short leg to keep his trunk vertical. That works, but only for a while. Eventually, the low back muscles tire from the extra work. Pain ensues. A shoe with a built-up sole and heel can compensate for the limb length discrepancy and make the spine happy, but not necessarily the owner.

We will learn later on what orthopedists can do when bones end up too long or too short, either from an injury or from a genetic condition, two categories of disease that can affect bone. What other categories of disease affect the world's best building material?

*Chapter 4*

# BONE'S OTHER FAILINGS AND WHO CAN HELP

IN THE PREVIOUS TWO CHAPTERS, I DESCRIBED how bones can go bad by not growing properly or by breaking. What else can happen to them? Let's think categorically, as medical students do when they comprehensively consider disease types, of which there are eight: congenital, traumatic, infectious, neoplastic, degenerative, vascular, metabolic/immune, and psychological. I do not know of any mental disorders that directly affect bone and I have already described fractures, so let's consider some examples of the other disease types that affect bones along with available treatments.

Conditions apparent from birth may have a genetic basis. An example is brittle bone disease, where a child sustains fractures merely by rolling over in bed or sneezing. It arises as a genetic mutation that causes either inadequate or faulty collagen production. Without the proper framework, the bones behave like sticks of chalk, which can snap with minimal force. Brittle bone disease comes in several forms. The most severe type results in multiple fractures to the skull, ribs, and limbs during delivery and is therefore lethal. Less severe forms are not lethal but problematic. Pediatric orthopedic surgeons insert rods into the central canal of the child's long bones to reinforce them before the fracture and stabilize them if they go ahead and break anyway. As the child grows, the rods must be replaced with longer ones, although there are new, inge-

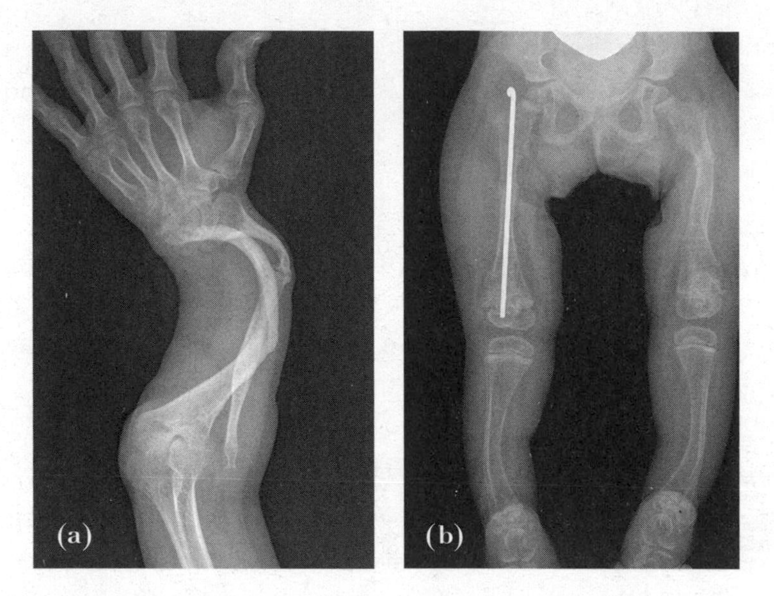


*(a) Brittle bone disease has caused multiple forearm fractures, which have healed with marked deformities. (b) A metal rod inserted into a bone's central cavity reduces the risk of fracture; and should one occur, the rod holds the bone in alignment.* (A) Benjamin Plotkin, MD; (B) Richard E. Bowen, MD.

nious devices that lengthen themselves when exposed to the force of an externally placed magnet.

In another genetic condition, the bones are normally strong but just short, and there are at least 200 forms of short stature. The most common type is caused by bones failing to elongate normally during growth. This means that the head and trunk are almost normally sized but with shortened limbs attached. (Welsh corgis and dachshunds are similarly affected.) Believe it or not, orthopedists have a treatment for children with this most common type of short stature, which I will describe later.

There are other conditions apparent at birth that do not have a genetic basis but are rather responses to some insult during pregnancy—alcohol, for example. Alcohol consumed by the mother during pregnancy easily passes through the placenta and wreaks all sorts of havoc on the fetus. Bones in the limbs can fuse together, but these are relatively minor issues in face of mental retardation and behavioral issues. Many other skeletal anomalies that develop during pregnancy have no known cause, yet they can be helped by orthopedic care. These include webbed fingers, absent thumbs, and clubfoot.

Bacteria are the cause of the next category of disease—infection. Bone is particularly susceptible to infection because its relatively sparse blood supply limits delivery of white cells, antibodies, and antibiotics to the crime scene. Intact skin, however, provides good protection. Simplistically, the body is just a bag of slightly salty water, and the bag (skin) protects us from drying out and from other environmental threats, including bacteria.

You may remember from eighth-grade health class or from Boy Scouts that the ends of broken bones at times may stick out through the skin, described as a compound fracture. This is a medical emergency, because despite all of bone's virtues, efficiently fighting infection is not among them.

Yes, bone is hard and compression resistant, which is at the core of this story, but these admirable features come at the expense of a rich blood supply. Enhanced nourishment would require numerous sizable holes in the bone to let arteries and veins pass through. After an injury to any tissue, blood delivers white cells and antibodies to stave off infection. The circulation also delivers oxygen and other nutrients to promote wound healing. Antibiotics arrive the same way. Hence, a good blood supply promotes efficient healing in all tissues.

Skin is awesome in this regard. It has a great blood supply as well as multiple other defenses against bacteria. Hence, skin lacerations heal readily. But poke a bone out through a hole in the skin, and bacteria attach to the bone and strive to infect it.

Sadly, bone does not have to be directly exposed to the outside world through a skin tear to suffer from bacterial attacks. These marauders can also arrive via blood pumped in from remote sites. As an example, tuberculosis germs residing in the lungs can enter the

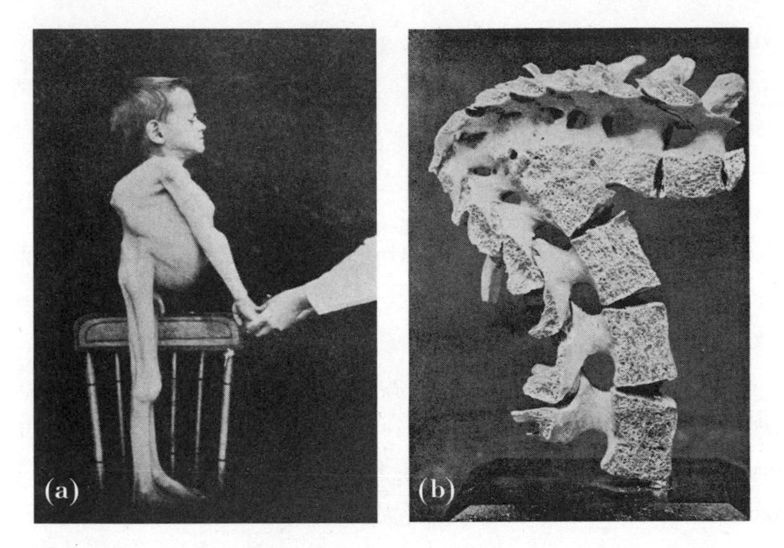


*Prior to the availability of antibiotics, tuberculosis could easily spread from the lungs and ravage bones and joints. Bacteria settling in the spine could cause collapse and deformity, as seen in these images from the early 1900s.* (A) ROBERT TUNSTALL TAYLOR, *ORTHOPAEDIC SURGERY FOR STUDENTS AND GENERAL PRACTITIONERS: PRELIMINARY CONSIDERATIONS AND DISEASES OF THE SPINE; 114 ORIGINAL ILLUSTRATIONS* (BALTIMORE: WILLIAMS & WILKINS, 1907); (B) WARREN ANATOMICAL MUSEUM.

bloodstream and find their way to bone, where they can settle and prosper. Shame on you, heart, for allowing this insult to happen to your loyal calcium banker.

Cartilage also is vulnerable to infection, since it has no blood supply at all and depends on joint fluid to nourish its cells. Bacteria in a joint, arriving either by direct access through an opening in the skin or via the bloodstream, can quickly destroy cartilage, leaving the underlying bones to grind away against each other.

As you know by now, bone consists principally of osteoblasts, osteocytes, and osteoclasts, and just like cells elsewhere in the body, bone cells can mutate and form tumors. Bone tumors are relatively rare, though, because even though the proportion of body weight that bones comprise is high (a roller bag's worth), bone cell turnover is relatively low compared with cells in the intestine, breast, prostate, and skin. As a result, there are fewer cell-division opportunities for bone cells to go awry. Bone tumors do occasionally occur, however, and can be either benign or malignant. Benign tumors have their origins in the bone itself and require treatment only if they cause pain, threaten the bone's strength, or bulge and interfere with function or appearance. A malignant tumor can either arise from bone cells or be transmitted to the skeleton from elsewhere in the body. To remember the common cancers that easily spread to bone, medical students use the mnemonic *BLT with a Kosher Pickle* (breast, lung, thyroid, kidney, prostate).

The next category of disease is degeneration. As people live longer, wear-and-tear joint changes, known as degenerative arthritis or osteoarthritis, have growing importance. Without undue provocation other than walking upright, the disease begins to cause some aching and stiffness in the spine by about age 40. What happens is that over time, the cartilage that covers joint surfaces tends to slowly wear away and may ultimately leave the underlying bones to rub against one another. Bone spurs then form at the margins of the joint, where in the fingers they are visible and perhaps cosmetically disturbing; but much worse, they can squeeze the nerves emerging from the spinal cord in the neck and low back. Pain ensues. The onset and progression of osteoarthritis is hastened in any joint that is not tracking properly or whose surfaces are not smooth for whatever reason.

Sometimes such misalignment occurs because the joint was just not properly formed to begin with. This can happen in infancy where the ball never seated completely in the hip joint's socket. By the time the person is 30, a painful limp develops. In other instances, a lax ligament lets a previously normal joint wobble out of alignment, which causes excessive cartilage wear. This commonly occurs in knees. Consider the analogy of driving a car with its front wheels out of alignment. The car still works but the tire tread wears away prematurely.

It should be clear by now that bone does not have much of a

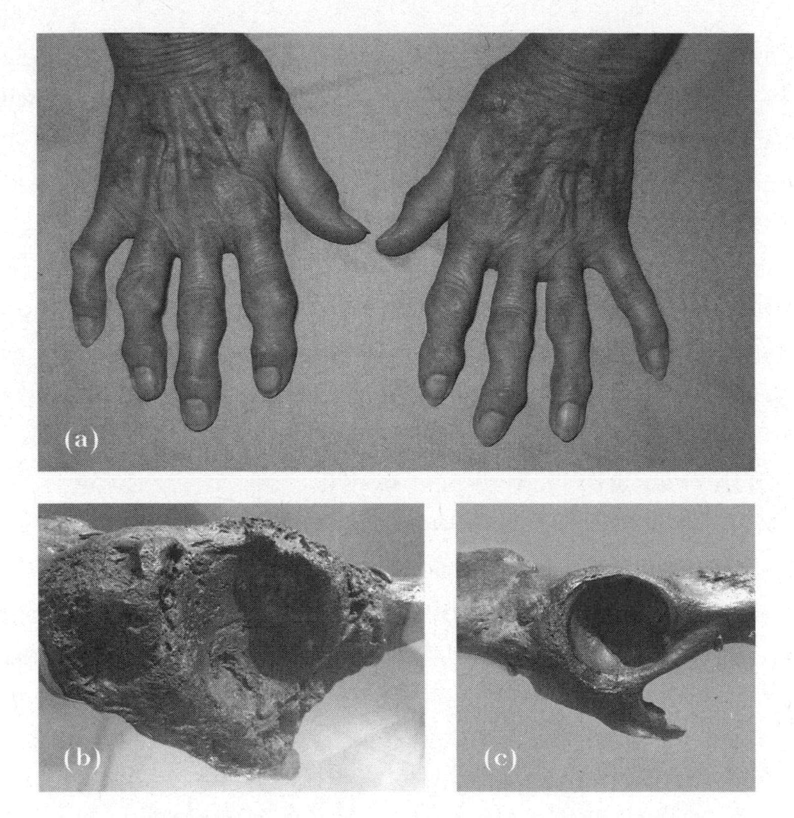

*(a) Bone spurs around the finger joints are the result of osteoarthritis and create their knobby appearance. (b) Osteoarthritis is not unique to humans, as evidenced by this saber-toothed tiger's hip socket. It is shallow and irregular with a markedly thickened rim. (c) This is especially apparent when compared with a normal hip socket in the same species.* (B, C) NATURAL HISTORY MUSEUM OF LOS ANGELES COUNTY.

blood supply and that its low demands are easily met under normal conditions. The heart, love it as we do, cannot pump blood to a site lacking the necessary arteries and veins. Loss of these essential blood vessels in bone can occur two ways—one obvious and one poorly understood.

A fracture tears adjacent blood vessels and pulls them away from the bone ends. A violent injury with wide fracture displacement will maximally strip away the blood supply. At times, the bone shatters into multiple fragments, some of which may be left with no blood supply at all. The art of fracture fixation includes achieving bone stability without further disrupting the bone's already tenuous blood supply.

Then sometimes for no clear reason, the blood supply to a bone, either totally or in part, can fold. The technical description is avascular necrosis—no-blood death. The loss of nourishment leads to the absence of cutting cones, which leads to weakened bone, which leads to collapse and deformity, which leads to arthritis. The hip is particularly vulnerable. When the head of this large weight-bearing ball-and-socket joint collapses, a painful limp ensues, which usually requires a total hip replacement for relief. Less frequent but equally irksome failures from an inadequate blood supply can occur in the wrist, shoulder, and ankle. Sometimes avascular necrosis happens entirely out of the blue. Other times and through an incompletely understood mechanism, high-dose steroids taken for severe asthma are causative.

To understand the next category of disease that affects bone, remember that the skeleton is the heart's calcium bank. Any time the pump gets a bit crampy, bone has to come up with a donation. In women, the contributions get bigger after menopause when estrogen levels fall, which weakens the bones and puts the hip and spine at particular risk for fractures.

For astronauts, the donations are huge even though nearly all the space explorers to date have either been premenopausal or not subject to menopause. Whiling away for months in the International Space Station, astronauts float around and cannot give their bones the opportunity to resist gravity. Without the stimulation from piezoelectric forces, the cutting cones go on vacation. Hence, orbiters lose calcium at 10 times the rate seen in earthbound postmenopausal women. On Earth or in orbit, this loss of calcium constitutes osteoporosis, which means porous bone. The depleted bone is fragile and prone to fracture. "You're welcome, heart, enjoy your precious calcium from inside your big strong protective chest cavity made of, all things, bone."

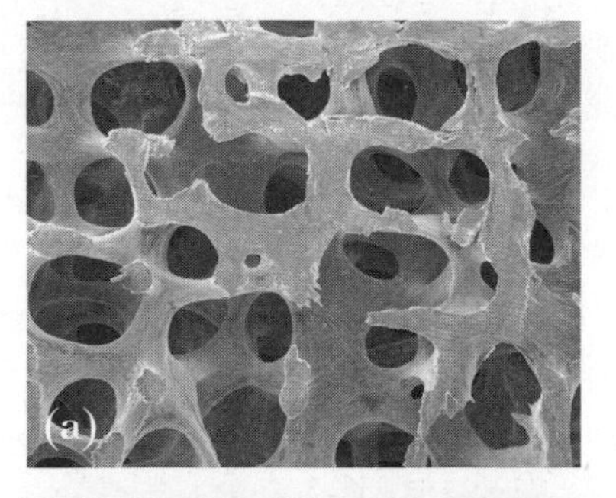


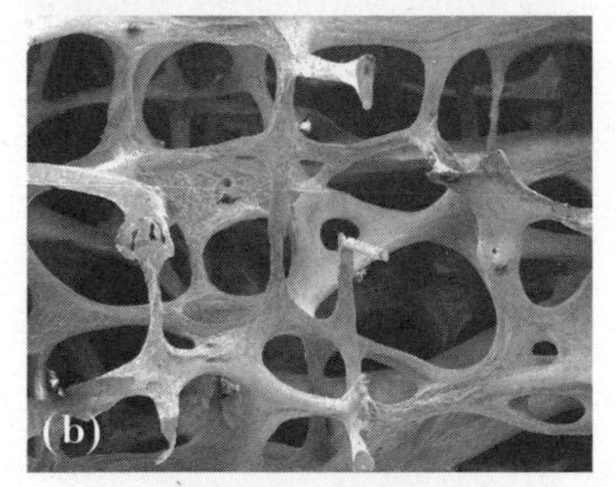


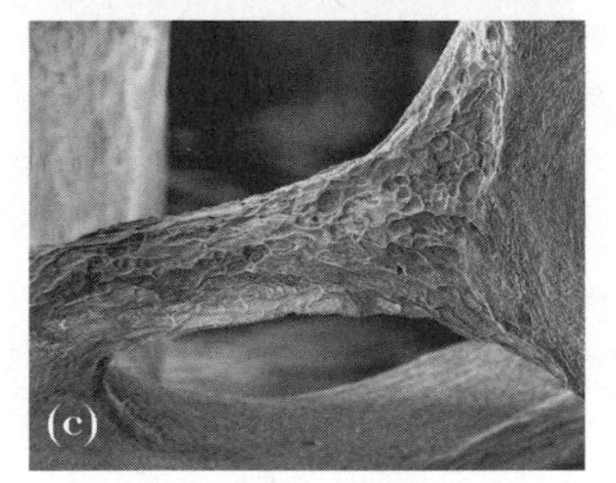


*These scanning electron micrographs show spongy bone in the low back. (a) Normal bony architecture has a sturdy, regular meshwork. (b) Hormonal changes and lack of mechanical stress cause thinning and weakening of spongy bone, findings indicative of osteoporosis. (c) This enlarged view of the osteoporotic bone shows craters and troughs created by bone-destroying osteoclasts.* Prof. Tim Arnett, University College London.

On Earth, frequent walking creates piezoelectric charges that stimulate the cutting cones to maintain bone density or at least retard its demise in the spine and hips. In orbit, the astronauts work out regularly with stretchy resistance bands to simulate gravity and stimulate piezoelectrically directed cutting cones. Although great for heart and lungs, bicycling on Earth or in orbit does not cut it for bone density. Those hydroxyapatite crystals have to sense some impact.

Along with mechanical stimulation, vitamin D and a hormone produced in the parathyroid glands are also critical to calcium balance and bone health. Vitamin D deficiency is one type of metabolic bone disease. The vitamin shortage can occur either from lack of sunshine-stimulated production in the skin or from nutritional inadequacy. Hormonal imbalance can also send bone into the skids. For instance, inadvertent surgical excision of the parathyroid glands during thyroidectomy can cause a parathyroid hormone deficiency. Smoking and possibly even air pollution also contribute to bone weakening.

In later chapters, I will describe treatments for some conditions in the congenital, traumatic, infectious, neoplastic, degenerative, and vascular categories of disease. Right here, however, is a good place to describe some types of medication used to prevent or treat osteoporosis, which is by far the most common bone disorder in the metabolic category. Please note that what follows is only a glimpse into the ever-improving treatment related to the complex chemical balancing act between bone production and turnover. My brief overview can enhance your appreciation of bone, but it is not meant to recommend any treatment; this can only come after a thorough assessment from and discussion with your doctor.

Medications for osteoporosis follow one of two strategies. First, there are the drugs that retard osteoclast activity and bone absorption. The largest class is the bisphosphonates. Don't let the word bog you down. The prefix "bis-" means "two," and these fairly simple chemical compounds contain two phosphorus atoms. Commonly encountered trade names include Actonel, Boniva, Fosamax, and Reclast. The first three are taken orally at either weekly or monthly intervals. Reclast is injected yearly into a vein. The bisphosphonates, whether they are swallowed and absorbed through the gut or given intravenously, have an affinity for hydroxyapatite. There the bisphosphonates inhibit osteoclasts from dissolving bone. Less osteoclast activity means more bone. Another medication, Prolia, works similarly to deactivate osteoclasts, although it is an entirely different and far more complex molecule. It is an antibody made of protein, so it costs more.

In older women, estrogen replacement therapy can simulate premenopausal life and reduce bone turnover, but its use increases the risk of breast and uterine cancers and is therefore not typically a judicious foil for osteoporosis. Evista is an artificial cousin of estrogen and works the same way estrogen does to maintain bone mass, but Evista does not apparently increase the risk of cancer. It can, however, cause other problems, including blood clots and hot flashes.

The other strategy to prevent and treat osteoporosis is to facilitate bone production. Forteo and Tymlos are synthetic forms of parathyroid hormone. With daily injections under the skin, they facilitate absorption of dietary calcium, which then becomes available for improving bone quality and keeping that pump happy.

An interesting approach for looking into disuse osteoporosis involves looking into caves. Bears hibernate all winter. That prolonged inactivity would render a human's skeleton mushy, yet bears seem to manage just fine. Brave investigators have darted grizzlies and similar species and sampled their bone and blood immediately before, during, and after hibernation. They found that a bear's bone metabolism essentially shuts down along with its other bodily functions during its winter rest. Osteoblasts, in the course of making new bone, normally send signals to the osteoclasts to remodel their creation. When the osteoblasts go inactive, so do the osteoclasts. Hibernation proves to be a highly complex, multisystem process affecting the brain, heart, kidneys, muscles, and bone. Investigators understand the process incompletely, so time will tell if bears can offer humans some help in preventing porous bone.

Rather than looking into caves, professionals who habitually look into our mouths have tried to determine if low-level fluoride treatment can prevent osteoporosis. Fluoride added to drinking water (1 part per million) hardens teeth and prevents decay. The same concentration seems to strengthen some bones and slightly weaken others. Higher levels of fluoride (4 parts per million) increase the skeleton's density but diminish its strength and resilience. Concentrations as high as 50 parts per million occur naturally in groundwater in some regions, especially near volcanoes, and drinking this water is universally disdained because it stains teeth and causes bone spur formation and spine stiffening. These facts speak to the complexity of bone metabolism and suggest that drinking nonvolcanic tap water is good while swallowing fluoridated toothpaste (1,000 parts per million) is bad.

Considering the multiple ways that bone can fail, from genetic miscoding to too much or too little fluoride, it should not be surprising that one group of medical specialists cannot oversee it all. We orthopedists like to think of ourselves as *the* bone doctors, but other specialists have integral roles, some by body region and others by disease type. Let's have a look. In deference to our scaly, furry, and feathery friends, remember that the same specializations for humans also tend to be available in veterinary practices.

Primary care physicians, consisting of internal medicine specialists, family practitioners, and gynecologists, provide the first line of defense in preventing and treating osteoporosis. Other medical specialists who have a stake in the nonoperative treatment of bone conditions include rheumatologists, endocrinologists, and physical medicine doctors.

Rheumatologists train in internal medicine and then take a two-year fellowship to learn diagnoses and nonoperative treatments for ailments affecting joints from head to toe. Your family doctor would likely send you to a rheumatologist if you had several persistently hot, swollen joints that could not be easily explained by injury.

Endocrinology is another subspecialty of internal medicine, again requiring a multiyear fellowship where one learns to diagnose and treat all of the maladies of endocrine glands, which include the pituitary, thyroid, parathyroid, pancreas, and adrenal glands, along with the ovaries and testes. To some extent, all of these glands have roles to play in bone growth and maintenance. The skeletal conditions that would most likely draw a patient to an endocrinologist would be those of calcium imbalance, especially osteoporosis and parathyroid disturbances.

Physical medicine and rehabilitation (PM and R) specialists, also known as physiatrists, are specially trained to offer nonoperative management of mechanical problems of muscles, bones, and joints. Included in their domain of expertise are back pain and rehabilitation after strokes and severe burns. And whereas rheumatologists and endocrinologists generally rely on medicines (both pills and shots) to help the body regain normalcy, physiatrists add braces, canes, walkers, and similar physical supports to facilitate function. They frequently supervise rehabilitation programs that are under the day-to-day care of non-MD specialists, including physical and occupational therapists, psychologists, and social workers.

Oncologists are cancer specialists and come in three forms. The first are internal medicine specialists with special training in hematology (blood disorders). Chemotherapy is their forte. They might treat blood cancers such as leukemia solely with medications and would be the ones to oversee a bone marrow transplant. Hematology-oncology specialists also combine efforts with radiation oncologists and surgical oncologists to give bone cancers a triple whammy. Radiation oncologists treat cancer with X-rays and radioactive compounds. Orthopedic surgery oncologists operate on bone tumors.

The word *pathology* combines the Greek words for "suffering" and "study of," and it follows that pathologists study disease. Even though pathologists do not *treat* disease, they are crucial members of the treatment team because they provide diagnoses based on microscopic and chemical analyses. Some pathologists specialize in diagnosing bone conditions. This requires particular patience because biopsies and surgical specimens of bone are way too hard to be immediately sliced thinly, stained, and examined microscopically. Rather, each specimen must be specially prepared for days to weeks before evaluation can begin.

Diagnostic radiologists look at bone indirectly via X-ray, ultrasound, radioisotope scanning, computed tomography, and magnetic resonance imaging. Interventional radiologists complete a normal five-year diagnostic radiology residency and then take an additional one- or two-year fellowship. Using fluoroscopic guidance, they can place needles into problematic areas of bones, such as into a painful vertebra collapsed from osteoporosis. An interventional radiologist can inject a liquid acrylic into the collapse site. The liquid hardens, relieving pain and preventing further bony collapse. Prior to injection, inflating a small balloon inside the vertebra may reduce or correct the collapse and improve the stooped-over position characteristic of the posture of a person with a severely osteoporotic spine. For certain bone cancers, interventional radiologists can place special needles that apply heat, cold, radio waves, laser light, or radiation to kill the malignant cells, all without the specialist directly seeing the bone.

How about doctors who have the profound privilege of seeing and touching living bone? This includes more specialties than you might guess. Starting at the head, plastic surgery and otolaryngology share the skull with neurosurgery. The first bone surgeries were likely performed by prehistoric neurosurgeons, since a number of skeletons unearthed from as far back as 8,500 years show 1-to-2-inch-diameter holes in the skull. Remarkably, the edges of the openings are smooth and rounded, which implies that these "patients" survived the drilling and that their osteoblasts and osteoclasts tried to heal the portal. This practice occurred in many different and widely dispersed cultures and before either writing or documented informed consent existed. Therefore, the purpose of these holes remains speculative. Perhaps they were ritualistic or were performed to relieve the pressure from indented skull fractures, which, if left unattended, would cause seizures or coma. Prehistoric medicine men (or women) may have recognized this and started drilling. In modern times, neurosurgeons may still access the brain by boring holes or cutting windows in the skull, but no longer with sharpened rocks.

The three bones in the middle ear are uniquely in the purview of the ear, nose, and throat doctors (aka ENTs or otolaryngologists). Although these bones count toward the total-body bone number of about 206, you could swallow a set without noticing, so they do not generate any great battles between surgical specialties regarding professional ownership of these bones.

The facial skeleton (cheek, eye socket, nose, and jaw) are managed by both plastic surgeons and otolaryngologists. Oral surgeons, who are specially trained dentists, also manage the tooth-containing bones.

Cardiac and thoracic surgeons share the breastbone and ribs, usually just needing to get past them to treat ailments of the heart and lungs. If you're unlucky enough to shatter your ribs or breastbone, which can occur from a hard fall or car accident, a general surgeon or thoracic surgeon can help. These injuries may be difficult to manage, not because of the fractures themselves but because the delicate, squishy contents of the chest cavity are directly adjacent and may have also sustained injury.

Spine surgery falls under the purview of both neurosurgeons and orthopedic surgeons. The relative interest of these two groups varies somewhat by area of the country and also according to the emphasis in each surgeon's residency training program. If you talk to one group, they may claim that they can manage it all—pinched nerves *and* bone and joint problems. If I were faced with spine surgery (having failed exhaustive nonoperative treatment overseen by a physiatrist), I would likely seek opinions from both an orthopedic spine specialist and a neurosurgery spine specialist. Then I would hope, especially for a complex problem, that one of each would be willing to collaborate on relieving any pressure on the nerves and stabilizing the bone to prevent further irritation.

Along with orthopedic surgeons, some plastic surgeons take an interest in hand and wrist problems, and podiatrists share interest in the foot and ankle with orthopedists. But when it comes to arms, legs, and pelvis, orthopedic surgeons pretty much have sole jurisdiction. Some say that orthopedists manage the bones that are too big to swallow. That makes orthopedists sound a bit beast-like, but considering the history of the specialty, that stereotype was not entirely inaccurate, as we will see.

*Chapter 5*

# BONE SURGERY THROUGH THE AGES

HERE IS A DISCLAIMER FOR TOTAL TRANSPARENCY. My name is Roy, and I am an orthopedic surgeon. I've been one for 40 years. I love working with and talking about bone. My peers do, too, as did our predecessors, who began scribbling about bone disorders and their management shortly after writing was invented. It is a rich and interesting history. Recounting some highlights is in order.

Imagine a heated prehistoric game of basketball where the two teams, the Skins and the Furs, use an inflated mammoth bladder for the ball. One player jams his finger and notes that it is pointing sideways. Instinctively, he yanks on it and successfully realigns the dislocation. Next week a teammate incurs the same injury, and the experienced one performs the same restorative maneuver. Over time he continues to learn from experience and achieves local acclaim as the go-to bonesetter. These skills are then passed down to his children. These bonesetters, along with shamans, midwives, and herbalists, developed in many cultures, including ancient Egypt and early Hawaii. Archaeologists have discovered Egyptian mummies with broken arms that bonesetters first splinted with strips of wood bark and then wrapped with linen. A papyrus from about 2900 BC records that such splints should be reinforced with plaster and honey. About 500 BC, Susruta in India and Hippocrates in Greece described stabilizing fractures with strips of wood, bamboo,

*As early as 2450 BC, Egyptians were treating forearm fractures (arrows) with splints made from bark. Blood-stained vegetable matter (*) indicates that the bone ends had penetrated the skin and caused some localized bleeding.*
G. ELLIOT SMITH, "THE MOST ANCIENT SPLINTS," *BRITISH MEDICAL JOURNAL* 1, NO. 2465 (MARCH 28, 1908): 732–34.

or sheet lead wrapped with string or linen strips stiffened with lard, wax, pitch, or egg white. Bandages saturated with blood and left to harden also sufficed.

Around 250 BC, the city of Alexandria, in Egypt, became civilization's center of scientific knowledge, and scholars came from great distances to learn. The Alexandrians shrewdly maintained their superior knowledge base by confiscating the visitors' learned writings as they entered the city. Officials had scribes copy the scrolls, placed the originals in the local library, and provided the travelers copies of their originals upon their departure. Alexandrians performed the first systematic human cadaver dissections. Visitors could also marvel at a human skeleton secured in proper alignment by fine wires and suspended vertically, another first that seems so commonplace today.

Beginning in ancient times, combat trauma has provided doctors with large numbers of injuries in healthy men in a short period of time. Again and again, this concentrated experience has added greatly to the understanding of wound healing. Such was the wisdom gained by the Greek physician Galen in about AD 150. He worked in Rome and was the gladiators' equivalent to a modern-day team physician. Galen therefore had the opportunity to manage gore, and he made many original observations about wound healing and wound management, some of which eventually turned out to be glaringly wrong. (Remember, he was the one who thought bone was made out of sperm.) Nonetheless, his writings were taken as dogma and prevailed in Europe for over a thousand years—truly the Dark Ages for medical advances.

Then the Enlightenment arrived. Ambroise Paré, a sixteenth-century physician and contemporary of the Renaissance artists and anatomists, served several French kings and their embattled soldiers. At the time, surgeons were using red-hot pokers or boiling oil to seal the raw, bleeding flesh of amputation wounds. Either treatment was agonizingly painful, and neither was particularly effective in staunching blood loss or preventing infection. Short of hot oil one day, Paré used threads from his coattail to tie off the ends of bleeding vessels before dressing the wound with the usual turpentine and gauze. Prompt healing ensued, and soldiers forever after have been grateful for improvements in surgery brought about by a battlefield necessity. Paré is appropriately known as the father of modern surgery.

Also during the Renaissance, physicians, who were university trained, considered surgery beneath them. They relegated operations—which consisted mainly of bloodletting and amputations—to the barbers, because although they were merely apprenticeship trained, they had the sharpest knives. Paré was among this guild of barber-surgeons. This meant that until the mid-sixteenth century, one could theoretically have a shave and an amputation at one sitting by the same practitioner. Thereafter, surgeons branched away from barbers and became separately chartered. Nonetheless, their work was demeaned.

Two holdovers from the barber-surgeon era persist. Red-and-white-striped barber poles represent the blood and bandages of the barber-surgeons; and in England, a surgeon is addressed as Mister whereas a physician is addressed as Doctor even though in recent centuries they have received the same core, university-based, medical education. British surgeons seem to be proud of the distinction and enjoy subtly flaunting their colorful history. Around the world, other medical specialists sometimes view surgeons as impulsive. We surgeons rather consider ourselves as decisive. Detractors have even described surgeons as being right or wrong but never in doubt.

The word *orthopedic* was coined in 1741 by Nicolas Andry, a French physician who wrote the first book on the topic, *Orthopédie. Ortho-* is Greek for "straight" or "correct," as in orthodoxy (correct belief) and orthodontics (straight teeth); and *pédie* stems from the Greek word meaning "child." In his book, Andry described how families and physicians could prevent and correct skeletal deformities in children. Of course, at the time the means of doing so were entirely nonsurgical. It would be another hundred years before general anesthesia and the thought of elective surgery came about. The graphic that Andry chose for the frontispiece of his book to illustrate his thoughts on straightening a child has become iconic.

Now skipping forward and across the Atlantic, Noah Webster published his monumental treatise *An American Dictionary of the English Language* in 1828. He aimed to simplify the old-world spellings of such words as *colour*, *cheque*, and *encyclopaedia*. Despite this learned lexicographer's best efforts, we still have two spellings for bone surgery: *orthopedic* and *orthopaedic*. Some stuffed shirts are reluctant to give up that "a" in *orthopaedic* because they say that *pedo* also means "foot," and these purists insist that *orthopaedic* means "straight children," which was Andry's intent, whereas *orthopedic* might imply just "straight feet." Somehow, American paediatricians long ago became pediatricians without apparent loss of professional standing. To my mind, Wikipedia brings the debate to an end. It says that *pedo-* relates to (1) children, (2) feet, (3) soil, and (4) flatulence. Or should it be flaetulence?

Nicolas Andry's publication of *Orthopédie* kickstarted orthopedics as a distinct specialty; and later in the eighteenth century, Jean-

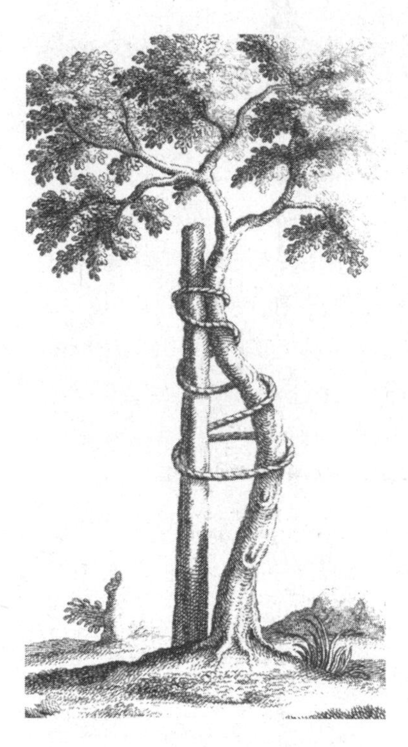

*This iconic drawing first appeared as the frontispiece in French physician Nicolas Andry's 1741 book* Orthopédie. *Just as a crooked tree's alignment can be improved during growth, Andry advocated similarly that bracing could prevent and correct childhood skeletal deformities.*

André Venel provided practical applications for many of Andry's teachings related to nonoperative treatments for children's foot and spine deformities.

During Andry's and Venel's time, surgeons were not subspecialized because there was nothing special that any one of them could do better than the others. This changed dramatically in the nineteenth century with the discovery of general anesthesia and with the gradual acceptance of Pasteur's proposal that bacteria were the source of infection. Up until that time, surgeons had no reason to wash their hands before surgery and would often wipe their surgical instruments on their coattails before replacing them in their kit. By mid-nineteenth century, general anesthesia allowed surgeons to operate more methodically and treat more complex problems. (Prior to this, a premium was placed on speed, and in one instance, the nearby fingers of the surgeon's assistant came off along with the patient's leg.)

A Dutch army surgeon named Antonius Mathijsen provided

another mid-nineteenth-century breakthrough that greatly eased the burden of applying casts to broken limbs. He applied plaster of paris powder to long strips of moist gauze and then rolled them up. When he needed a cast, he momentarily immersed a roll in water to activate the plaster of paris and then circled the injured limb multiple times with the bandage, which quickly hardened. Undoubtedly the solidified plaster cast smelled better than the previously recommended lard, egg white, or old blood. Several years later, Mathijsen's invention received a workout during the Crimean War. Legend has it that when military surgeons during that conflict were short of water, they would moisten the plaster of paris with urine. The Crimean War also made Florence Nightingale famous for her achievements in organizing nursing care for wounded soldiers, probably including those with reeking casts.

Around the same time there was also growing awareness and acceptance of the germ theory of disease, which slowly led to the development of aseptic techniques and the use of rubber gloves and surgical drapes. After that, operations could last hours with good prospects that the patient would not only survive the ordeal but eventually heal without infection.

All these pioneers just described were medical doctors, MDs, with the roots of the discipline going back to Hippocrates about 2,500 years ago. A branch discipline developed far more recently and appropriately did so in the Show-Me State. DO is short for doctor of osteopathic medicine. *Osteo*, of course, means "bone," and *pathy* means "disease." Missourian Andrew Still, MD, introduced osteopathic medicine in the late 1800s. His research and observations led him to an understanding that the musculoskeletal system was central to health and disease throughout the body. He believed that improving the body's structure through the application of manual techniques, now known as osteopathic manipulative medicine, would promote normal function and self-healing of various organ systems, including digestive and respiratory ailments. At the time, many conventional medical treatments were not particularly effective, so osteopathy quickly developed a following. Dr. Still opened the first school of osteopathy in 1892.

Reflective of Dr. Still's holistic philosophy, many DOs traditionally gravitated to primary care—family practice, general internal medicine, and pediatrics. Nonetheless, graduates from schools of osteopathic medicine today may choose to specialize and can do so either in osteopathic or in MD residency training programs. In most communities, there are far more MDs than DOs, but as both trainees and practitioners, MDs and DOs often work side by side, sharing medical ownership of bone.

At the time when Dr. Still was developing osteopathy, some MD surgeons had already begun to specialize in treatments of the brain, eye, or other body parts; but fracture treatment remained within the domain of the generalist in cities and at times left to the care of bonesetters in rural and impoverished areas. This changed with the Industrial Revolution and specifically with the building of the Manchester Ship Canal in England, which remains the world's longest river navigation canal at 36 miles. Rather than wartime, when rapid advances in the treatment of trauma typically take place, this was a massive peacetime project involving hundreds of cranes, locomotives, and excavators, thousands of trucks and wagons, and 17,000 construction workers. The mix produced a great number of skeletal injuries over the six-year construction period.

A few years earlier and because of hard times at home in London, Robert Jones, then a teenager, moved to Liverpool to live with his uncle, Hugh Owen Thomas. Thomas was an orthopedist. His father, grandfather, and great-grandfather had been bonesetters. Thomas made multiple contributions to the management of skeletal diseases that included publishing treatises on tuberculosis and on femur fractures. He encouraged his nephew, Robert, to attend medical school and then to join him in practice, which Robert did. Together, Thomas and Jones developed a special interest in fracture management, whereas most orthopedists at the time dealt primarily with children's skeletal deformities.

In a fortuitous turn of events in 1888, Jones became surgeon-superintendent of the Manchester Ship Canal construction project and took advantage of this opportunity to develop the first comprehensive accident service in the world. He spaced three hospitals at intervals along the canal with first aid stations interspersed. Jones staffed the hospitals with personnel skilled in fracture management and operated on many of the injured workers himself. This intense experience of operative as well as nonoperative fracture management contributed to the development of techniques for improved fracture care. The newly gained knowledge also proved invaluable during the soon-to-follow First World War, and Jones was apponted Inspector of Military Orthopaedics, overseeing a 30,000-bed organization.

Thomas invented a splint for temporary immobilization of broken legs, and Jones devised a bulky bandage to be used after knee surgery. Both of these advances bear their innovator's name and are still used today. The most noteworthy and lasting mark that these two orthopedists made on medicine, however, was that they defined the specialty. Canal building and then war—these experiences with skeletal injuries, both intensive and extensive, ended a decades-long discussion about the nature of the specialty of orthopedics. Should it include surgical procedures or just focus on straightening crooked children with casts and splints? Since 1920, everyone appropriately calls the specialty orthopedic surgery.

Early in the twentieth century, all orthopedic surgeons were men, often large men with big hands. They had an advantage because relocating a dislocated hip and hammering, sawing, and drilling hard bone by hand was physically demanding. Somebody, likely intimidated by an early orthopedist's physical bulk, characterized us this way: "They are as strong as oxen and twice as smart."

What sort of a medical student is attracted to this upbeat realm of orthopedic surgery? Certain stereotypes exist for all specialties in medicine, and of course exceptions abound, but many medical students are drawn to orthopedics because they sustained playground or sports-related injuries themselves. The orthopedist's intervention that returned them to their game led them to think, "Hey, I could do that," and some jocks become orthopedists. Some became famous for their accomplishments either as Olympians or professional athletes. These include Mark Adickes (football), Eric Heiden (speed skating), Alec Kessler (basketball), Dot Richardson (softball), and Jason Smith (hockey).

Other medical students, myself included, grew up using tools in the workshop or garage. When we discovered that we could use sterile versions of many of these same tools in the operating room, we were excited to get going.

Perhaps, too, medical students come to realize that patients only rarely die from musculoskeletal diseases. Furthermore, orthopedists typically manage quality-of-life issues rather than death-or-life issues. This appeals to some and not to others. Those who need to "hold life in their hands" may choose to become neurosurgeons or cardiac surgeons. Orthopedists would rather be happy and get patients back in their games, both returning athletes to their sports and restoring an oldster's ability to rise from a chair comfortably.

Then something else happens, whether or not a medical student is consciously aware of it. During the third year of medical school, each student becomes a junior member of various treatment teams. In turn, they live the day-to-day hospital lives of pediatricians, psychiatrists, internal medicine specialists, surgeons, and obstetrician-gynecologists. On their surgery rotation, they get a taste of general surgery and several surgical specialties, including orthopedics, plastic surgery, urology, and neurosurgery. They may find that they love or hate making rounds early, staying late, standing at the operating table in the middle of the night. On a gut level, medical students discover which group of specialists they like being around. This includes the way they think, the problems they confront, the treatment methods they use, how they interact with other team members, what makes them laugh, and what they do in their free time. Perhaps unconsciously, medical students select a specialty where they know they can succeed, where they can contribute and feel good.

Even aware that no generalization is worth a damn, here is my characterization of your friendly bone doctor: happy, optimistic, result oriented, energetic, efficient, hardworking, decisive, and gregarious. Also, orthopedists excel at the task of grasping three-dimensional relationships. For instance, when fixing a hip fracture, we look at front-to-back and side-to-side images of a broken hip on the operating room fluoroscopy unit. Then peeking only at the outside border of the thighbone through a small incision, we drill a 4-inch-deep hole into the round head of the thighbone and insert a long screw. This is somewhat analogous to standing at your front door and shooting an arrow into the upstairs bathroom light fixture aided only by a floor plan. I think that this trait is innate. Those who are good at interpreting flat images into three-dimensional reality might be attracted to orthopedics, interventional cardiology, or interventional radiology. Somebody who cannot read a map or walk safely through their house in the dark would probably find more happiness in less spatial-relationship-demanding specialties such as pediatrics or internal medicine. (I am not disparaging other specialties. They require traits that I do not have.) Furthermore, various studies have shown that left-handed individuals are better at spatial orientation than righties. It is my (entirely unconfirmed) impression that perhaps 20 percent of orthopedic surgeons are left-handed compared with approximately 10 percent in the general population. It seems to go hand in hand, so to speak—spatial abilities and a lefty's attraction to orthopedics.

Another aspect of an orthopedic surgeon's life makes it both emotionally rewarding yet stressful. Joint motion along with limb alignment and length on one side of the body are easily compared with that on the opposite side. When things go well, patients express gratitude to have function and appearance restored. If anything is slightly off, however, the result may be forever evident to even casual observers. Contrast this to a lung or bladder that is not working precisely right when treatment is complete. The treating doctor may not be roundly praised, but because the results are not in public view, she is not likely to be severely condemned.

Speaking of she, these days there is nothing about the physical demands of orthopedic surgery that precludes women from practicing orthopedics competently. The time of orthopedists being characterized as Attila the Hun's offspring are long gone. Power drills and saws along with assistants and devices to help position patients have taken brute force out of the equation. What remains, however, are the irregular hours dictated by emergency room coverage; and this lifestyle issue may deter some from the specialty, men as well as women, who presently constitute approximately 7 percent of practicing orthopedists. This is slowly changing, however, because in recent years women account for nearly 18 percent of full-time orthopedic faculty, and 14 percent of orthopedic surgery residents are female.

Specialists of all sorts enjoy a little stress-relieving hospital humor, and all of us may poke fun at one another. This aphorism takes a broad swipe: "The internist knows everything and does nothing; the surgeon knows nothing and does everything; the psychiatrist knows nothing and does nothing; and the pathologist knows everything and does everything, just a week late." Surgeons would not entirely disagree about their characterization and might repeat the adage, "A chance to cut is a chance to cure." Others have accused orthopedists (who, remember, are the ones who treat bones too big to swallow) of handling (or ignoring) problems in the operating room by exhorting, "If you can't hear it, it's not bleeding" and "If it doesn't go easily, force it."

Are orthopedic surgeons as brutish as these well-worn jabs suggest, or are other specialists just sniping out of jealousy? If they are surgeons, perhaps their jealousy stems from awareness that bone heals without scarring, while the brain, liver, lung, bladder, and other tissues in which they have special interest permanently scar when cut. If the snivelers are not surgeons, perhaps their envy relates to awareness that patients with musculoskeletal problems usually get better and move on, since orthopedic treatments may entirely restore a patient's quality of life. Not so for patients with diabetes, emphysema, or psoriasis. For these conditions, the doctor may only be able to palliate the problem. Of course, I am glad that there are specialists that have the temperament and skills to deal with such chronic conditions, but I like fixing things completely.

If an orthopedic surgeon no longer has to be a hulking giant with mitts for hands, what does it take today to join the club? For starters, it takes a bachelor's degree plus at least nine more years and a certain level of smarts. After the first two years of medical school, all US medical students take Step One of the United States Medical Licensing Examination (USMLE). This probes their knowledge in anatomy, behavioral science, biochemistry, microbiology, pathology, pharmacology, and physiology and touches on nutrition, genetics, and aging. You might try pronouncing USMLE as "You Smile," but second-year med students more likely consider it "Big Ulcer," because to a great extent their score opens or closes opportunity's door for their selection of a medical specialty.

Residency directors use the USMLE score as a screening tool to determine which applications to scrutinize and to decide which applicants get invitations to interview. Students with mediocre scores are relegated to applying for training in less competitive specialties. In recent years, otolaryngology, dermatology, orthopedic surgery, and plastic surgery have led the list of 21 residencies as being the most competitive. Certainly there are students with top USMLE scores who choose to enter less competitive residencies, but those with average scores will likely be closed out from taking care of skin, throats, or bone.

Immediately on completion of medical school, students take USMLE Step Two, which tests knowledge in surgery, medicine, pediatrics, and the other clinical specialties. Step Three comes during the first year of residency and examines the in-the-trenches clinical application of book learning that the first two steps tested. If trainees pass Step Three and complete their first year of residency, aka internship, they can get a state license and hang out their shingle. In times gone by, that is what many new doctors did. They literally started their practice by, well, practicing.

Most medical school graduates these days choose residency training and develop special expertise. These formal programs began in the late nineteenth century. To be close to the action and to minimize the financial burden on both the institution and themselves, these newly minted doctors quartered in the hospital under rather monastic conditions. They literally were residents and sometimes continued as such for an indeterminate number of years until their chief declared them sufficiently well trained and let them loose.

Today, orthopedic surgery residency lasts five years. During this time, trainees are exposed to all of the subspecialties of orthopedic surgery. They assume increasing responsibilities for patient management as they develop their knowledge, skills, and judgment. Trainees start out being unconsciously incompetent. ("Hey, I've never tried that, but it looks easy.") With some humbling experience, they become consciously incompetent. ("It's not as easy as it looks.") Conscious competence ensues. ("I can do this by carefully going step by step.") Then, perhaps after years in practice, they gradually become unconsciously competent. ("It's easy, I just do it.")

The residents learn a lot from one another, especially from the trainees who are just a year further along, because the steps necessary to acquire a new skill are still fresh in the head of a consciously competent teacher. For fun, we sometimes oversimplify the residency experience by describing it as see one, do one, teach one.

It may put your mind at ease to know that not all of the seeing, doing, and teaching occurs in the operating room. In a surgical skills lab, residents hone their techniques on plastic skeletal models that closely simulate real-life conditions. A thorough understanding of anatomy is the mainstay of all surgery, so trainees also spend time dissecting cadavers to solidify their knowledge of musculoskeletal anatomy and to enhance their dissecting skills. All of us owe a debt of gratitude to those who will their bodies to medical schools.

In addition, residency programs require that all trainees perform some research during these years. This helps advance the specialty and also exposes the trainees to experimental methods and rigorous critical thinking. If they never again present or publish another paper, this firsthand exposure to research forever allows them to easily extract the true value of other investigations.

Each year in residency brings another standardized test, the Orthopaedic In-Training Examination (OITE), which is a comprehensive multiple-choice test whose results include placing each examinee in a national percentile ranking with all 700 orthopedic residents in the same year of training. Residency directors use the OITE results to ensure that each trainee is progressing satisfactorily, and if not, to take remedial action that infrequently even includes dismissal.

In recent years, nearly all orthopedic residency graduates extend their training another year to take a subspecialty fellowship. This often involves moving to another institution. Orthopedic fellowships are available in hand, shoulder and elbow, foot and ankle, spine, oncology (tumors), pediatrics, sports medicine, trauma, and total joint reconstruction. During this year, a fellow works with one or more "wizards" in the field and develops knowledge and skills needed to care for the most complicated conditions managed by that subspecialty.

To become board certified, a credential demanded by most hospitals, the green orthopedic surgeon takes a two-part examination, which the American Board of Orthopaedic Surgery administers. Whereas the first part is another multiple-choice test, the second part, taken after two years in practice, consists of four 25-minute oral examinations, each before two examiners who delve into the detailed records of 12 patients whom the applicant has treated. The applicant brings all medical and billing records and imaging studies for the examiners to scrutinize. If successful, the beleaguered young orthopedist is now at least 11 years out of college and has passed at least 10 examinations.

So that's it for examinations, right? Wrong. Every 10 years in practice, board-certified orthopedic surgeons (and most other specialists as well) have to demonstrate maintenance of competency to their certifying board. It is a good thing that bone doctors enjoy their work and are likely to willingly stay current and competent. Passing these checkups demonstrates to the public that this is so.

Residents always ponder where they should hang their shingle after completing their training. Various studies have shown that an orthopedist can manage the musculoskeletal ailments of 17,000 to 20,000 people, more if physician assistants and nurse practitioners are available to help. To look at it the other way, a catchment area of 20,000 people can support a bone doctor. By comparison, the same area needs 10 times as many primary care physicians. Not every young practitioner follows the aphorism "Go where you are needed." San Diego and other lovely cities are saturated with specialists of all sorts, whereas less desirable locales are often underserved. The distribution tends to smooth out, however, with trial and error. Half of young orthopedic practitioners change job positions within two years of completing training. Sometimes even when located in out-of-the-way locations, a few will rise to international visibility and become giants.

*Chapter 6*

# SIX ORTHOPEDIC GIANTS

In this chapter I'll introduce six orthopedists who followed the advice about going where they were needed, found themselves in unique settings, and made original, monumental contributions toward advancing the treatment of bone diseases. I have already highlighted Hugh Owen Thomas and his nephew Robert Jones. Their work in the late 1800s helped expand orthopedics beyond its traditional scope of amputating injured and infected limbs and treating deformities and weaknesses caused by rickets and polio. Jones was knighted for his contributions.

Also in Britain and also eventually knighted, John Charnley pioneered total hip replacement in the mid-twentieth century. Up to this point in human history, many people living into their 60s and 70s developed wear-and-tear degenerative joint disease—osteoarthritis. When this occurs in fingers, the joints become stiff and knobby and perhaps painful. With a few finger joints affected, people manage by transferring tasks to other digits. The state of affairs is quite different, however, when the disease affects the hips and grinds away the hip's smooth gliding cartilage, which leaves the bone surfaces to rub on one another. Walking becomes painful. Climbing stairs or rising from a chair is worse. For millennia, a cane, crutches, or wheelchair was the only remedy short of surgery, which was primitive and only marginally effective.

The original surgical treatment was simply to cut off the knobby end of the thighbone where it ground against the pelvis. This reduced pain but rendered the limb short and unstable. Veterinarians today perform the same operation on dogs with arthritic hips, but because our canine friends share weight-bearing and ambulation among four limbs rather than two, they do better.

Following the discovery of general anesthesia in 1840, surgeons began interposing various materials between the hip's worn-out surfaces. Their ingenuity knew no bounds, and they were not burdened by any concept of informed consent, a twentieth-century development. Sure, the doctors followed the Hippocratic teaching of "first do no harm," but who was to know how various materials implanted between the surfaces of an arthritic joint would fare unless somebody tried it? And try they did, using fat, muscle, pig bladder, celluloid, wax, glass, rubber, and sheets of zinc, magnesium, or silver. Robert Jones, of Manchester Ship Canal fame, even tried gold foil.

Placing a thin sheet of some material between the opposing and arthritic joint surfaces posed several problems. First, the body had to tolerate the interposition and not reject it. Then the procedure needed to restore at least some motion and reduce the patient's pain. All early efforts at resurfacing failed, which led investigators to start replacing rather than relining the joint. First they tried rubber, then ivory, then various metallic ball-and-socket components, but complete joint replacement posed a new problem. The artificial joint components required secure fixation to the patient's pelvis and thighbone in order to prevent any wobbling or pistoning between the implants and the skeletal areas where they were seated. In 1891, Berliner Themistocles Gluck first tried securing ivory implants with metal screws and then turned to mixtures of plaster of paris, powdered pumice, and resin for fixation. All failed.

In the first half of the twentieth century, multiple European and American orthopedic innovators experimented with different fixation techniques and with variously configured total hip components made from different metal alloys. It was Englishman John Charnley who brought total hip replacement into practical and widespread use. Early in his medical career, Charnley fell under the influence of orthopedists in the Manchester area whom Robert Jones had trained. At this time, orthopedics was still considered a minor specialty and did not initially attract the interest of Charnley's fertile and inventive mind. Duty in the Royal Army Medical Corps during World War II took Charnley to Egypt, where he collaborated with engineers in developing various orthopedic braces and surgical instruments. On return to civilian life, Charnley focused his research on bone grafting, fracture healing, and ultimately on the weighty problem of total hip replacement. An offhand comment in the 1950s got him going: a patient who had received an artificial replacement commented to Charnley that every time he leaned forward at the dinner table, this hip squeaked loudly, which greatly disturbed his wife and prompted her to leave the room. This set Charnley off on a lifelong quest to perfect a *low-friction* total hip replacement—one that would not squeak and, by implication, one that would glide smoothly and not wear out.

Twice along the way, his curiosity and inventiveness led him to experiment on himself, first by having a colleague place an experimental bone graft in Charnley's leg. Infection ensued and required several further operations before healing occurred. Later, Charnley injected wear debris from one of his total hip designs into his own thigh to observe the resultant inflammatory response.

Charnley machined some of his early components at home on a lathe, which he had purchased using royalties gained from previous inventions. By the late 1950s, he had tried various hard plastics and settled on PTFE (Teflon) for the socket side of the joint. PTFE did not squeak and Charnley enthusiastically implanted it into 300 patients before its shortcomings became clear. Why did it take that many? The PTFE was suitably durable the first year, less so in the second, and not so in the third, when wear debris became problematic. The body recognized the debris particles produced by hip motion as foreign agents and fought them off with inflammation—pain, heat, swelling, redness. The only recourse was to remove the artificial joint. Charnley was distraught, although his patients were not. They remarked that the surgery had given them several years of pain relief.

In 1962, a salesman stopped by the hospital selling plastic gears, which came from Germany and which the weaving industry was beginning to embrace. He left a chunk of this little-known, specialized polyethylene, quite hard and dense, with the hospital's supply officer, who passed it on to Charnley's lab director. The investigator began immediately to test its wear properties even though Charnley's initial response, after digging his thumb into it, was that the lab director was wasting his time. Nonetheless, after 3 weeks of around-the-clock testing, the material showed less wear than the PTFE did in one day. Charnley noted later, "We were on."

Two conundrums remained. The first was the need to solve the problem that Gluck had approached decades previously—how to secure the artificial components to the bones such that the implants would not loosen. Glues would not stick to wet surfaces. Dentists, however, had successfully solved a similar problem by using acrylic cement to secure dental implants into jawbones, and Charnley recognized its potential for total hip arthroplasty. At surgery, the scrub tech mixed powder and liquid forms of the acrylic into a creamy paste, which Charnley used to cover the prepared bone surfaces before positioning the total joint components. The acrylic filled in any irregularities between the bone and the implants and hardened within minutes. This "grout" allowed wide and even distribution of forces back and forth between the components and the bones to which they were attached. It worked and is still used today. By the way, the same acrylic shows up in manufactured products under the trade names of Plexiglas and Lucite.

The other problem that hounded Charnley was bacteria. Implanting these sizable metal and polyethylene components through large incisions greatly risked wound infections, which necessitated removing the components and leaving the patients no better off than if they merely had the thighbone's knob sawed off originally. Bacteria

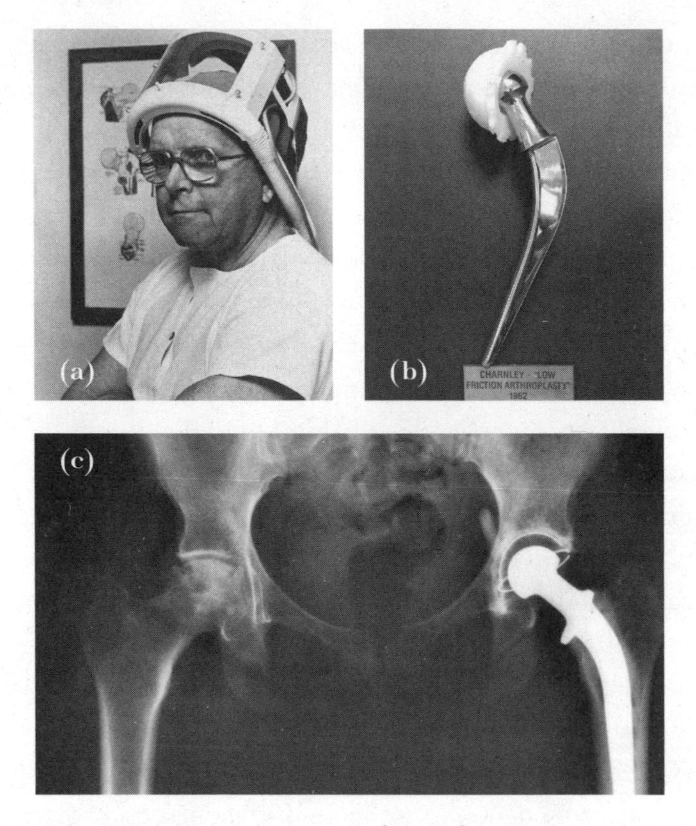


*(a) Sir John Charnley has the hood of his infection-reducing "space suit" raised between surgical cases. (b) An artificial hip consists of a polyethylene cup, which is secured to the pelvis, and a metallic ball-and-stem component, which is inserted into the thighbone. (c) This X-ray shows the positioning of the new hip components with a wire marker outlining the cup. The design of the pictured implant is one of many that have followed on the success of Charnley's original design.* (A) NATIONAL LIBRARY OF MEDICINE; (B) TOM SCHMALZRIED, MD; (C) NATIONAL INSTITUTE OF ARTHRITIS, DIABETES, DIGESTIVE AND KIDNEY DISEASES.

inevitably floated on air in the operating room, and some were likely drifting into the wound or settling on the surgical instruments, which then transferred them into the wound. Charnley mounted a three-pronged attack. He devised "space suits" to be worn by all of the operating room personnel. Each outfit had a ventilation system, which kept the users comfortable even though they were isolated from head to toe from the air circulating in the operating room. Then Charnley arranged for the room's ventilation system to include highly efficient filters that cleansed the air entering the room of most bacteria and dust particles. Furthermore, the filtered air was channeled into a "laminar flow" pattern such that the cleanest air in the room was always in the vicinity of the surgical incision.

By the late 1960s, Charnley had worked out the problems just discussed, and total hip replacement became practical and safe. Orthopedic surgeons worldwide beat a path to his door to learn how to do the procedure. Before they could return home with a set of Charnley's instruments, he insisted that they take his two-and-a-half-day course covering his methods, not only to be able to reproduce Charnley's operative techniques and results but also to mirror his meticulous recordkeeping needed for future analyses.

The composition and shape of the total hip implants continue to undergo refinement, as do the surgical techniques for implantation. Today, over 300,000 Americans receive total hips annually, and nearly 1 percent of Americans have at least one. With respect to quality-of-life improvements, total hip arthroplasty compares favorably to the benefits bestowed by the treatment of high blood pressure with medication, chronic kidney failure by dialysis, and coronary artery disease by stents and bypass grafts.

Other orthopedic surgeons have extended Charnley's work to develop similarly successful replacements for arthritic knees and shoulders. When individuals are otherwise healthy, a severely arthritic and painful large joint can be debilitating, reminding them of their frailty with every movement. Charnley's contributions to modern medicine were monumental and resulted in knighthood.

Whereas Sir John was recognized and celebrated during his lifetime, other orthopedic pioneers have not been so fortunate, a phenomenon not unique to orthopedics. Throughout history, contemporaries have a habit of ignoring, discounting, or demeaning original thinkers. Galileo was condemned to house arrest for posit-

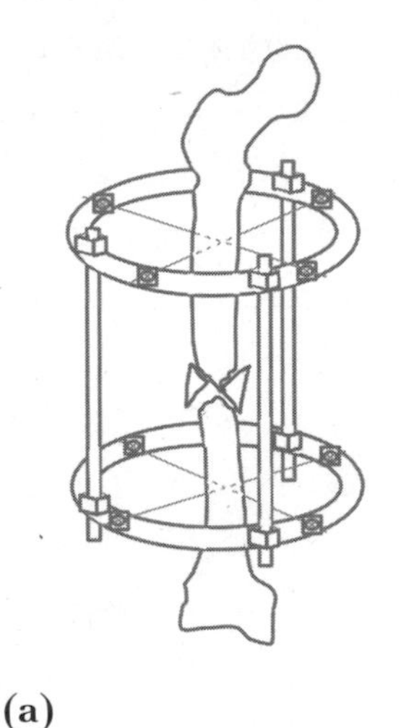

*(a) This external frame, secured originally to a severely injured bone with bicycle spokes, revolutionized treatment of complex fractures that had failed conventional treatments. (b) The inventor, Dr. Gavriil Ilizarov, left, is walking with his famous patient, Valeriy Brumel, Olympic high-jump gold medalist. Brumel has an Ilizarov frame on his right leg.* (B) SVETLANA ILIZAROV, MD.

ing that Earth was not the center of the universe. I do not know of any orthopedic innovators who were placed on house arrest, but I will introduce several pioneers whose road to fame and glory in their own time was circuitous and rocky if not altogether absent.

The first is Gavriil Ilizarov, a Pole. He attended medical school in Crimea and Kazakhstan during World War II and then, without any practical experience, accepted a post in Kurgan, Siberia. This war-torn region was 1,200 miles east of Moscow, far away from any established center of advanced medical understanding. The area was rife with wounded soldiers suffering from infected fractures that would not heal. With vast need, limited resources, and no preconceptions to restrain him, Ilizarov developed an external fixation frame, which would support a shinbone or thighbone fracture during healing. As others had done before, he placed pins perpendicular to the bone on both sides of the fracture site and left the pins protruding well outside the skin. He attached the pins to metal rings that surrounded the limb, several above the injury site and several below. Ilizarov then completed the external fixator by fastening all the rings together with longitudinally aligned metal bars. His device differed from forerunners because Ilizarov used threaded rods for the longitudinally oriented struts.

By 1955, Ilizarov had become chief of trauma and orthopedics at his Siberian outpost. Scarce resources required improvisation. He used bicycle spokes for the bone-penetrating pins. Ilizarov compared the resulting construct to bicycle wheels, where the bone ends were the hubs, which were fully stabilized by spokes passing from the hubs to the rims.

By eliminating any motion between the bone ends brought into contact by the fixator, osteoblasts could finally heal the fracture. Bone gaps, however, were problematic because osteoblasts can jump only so far, across ditches but not canyons. To close gaps, Ilizarov slowly moved the rings toward each other by using a wrench to make daily, tiny adjustments of the rings' locations on the threaded rods.

For patients whose fractures manifested gaps, Ilizarov showed them how to perform these serial adjustments at home. It took weeks. One confused patient persistently turned the wrench the wrong way, gradually widening the fracture gap rather than closing it. To Ilizarov's surprise, the expanded gap had formed so slowly that the bone healed anyway. It did so in the usual manner via osteoblasts producing collagen and hydroxyapatite. These microscopic workers toiled happily, unaware that their task was expanding.

Other surgeons had lengthened limbs by external distraction before, but they had always filled the gap in the lengthened bone with bone graft taken from elsewhere in the patient's body. This necessitated additional surgery to harvest the graft and risked the development of donor site pain, disfigurement, and disability. In an aha moment, Ilizarov realized that by moving the bone ends apart ever so slowly (six adjustments per day to produce not even a sixteenth of an inch of total movement), new bone would fill in the gap on its own. (Yank on a licorice stick, it breaks. Pull slowly, it stretches.) Furthermore, Ilizarov recognized that the technique could correct bones that had healed in any combination of shortening, angulation, and rotatation. (Twist and bend a licorice stick slowly, it twists and bends.)

Ilizarov applied the method widely, and his patients called him "the magician from Kurgan." Nonetheless, the medical establishment in Moscow considered him a quack and discounted his growing achievements and reputation. This began to change when Russian high jumper Valeriy Brumel broke his leg in a motorcycle accident in 1965, a year after winning the Olympic gold medal. Following three years of multiple and unsuccessful operations in Moscow to heal the injury, Brumel traveled to Kurgan and placed himself under Ilizarov's care. The athlete recovered sufficiently to high-jump an awesome 6 feet 9 inches, which was 7 inches off his world record but still quite respectable for somebody who had been hobbled by injury for years.

Regardless of his success in treating Brumel, Ilizarov's contributions still did not receive the recognition they deserved—even though his center in the 1970s grew to 24 operating rooms, 168 physicians, and around 1,000 beds, making it by far the largest orthopedic center in the world.

Then in 1980, an Italian adventurer sought Ilizarov's help after European doctors had given up hope of ever producing a sound leg. The mountaineer had broken it 10 years earlier and was left with an unhealed fracture and an inch of shortening. After Ilizarov achieved bone healing and lengthening, the grateful patient called Ilizarov "the Michelangelo of Orthopedics." On return to Europe, the patient's result astounded the Italian doctors, who then invited Ilizarov to speak at a European fracture conference in 1981. Ilizarov gave three lectures, the first time he had presented his material outside the Soviet Union. At the end he received a 10-minute standing ovation.

In subsequent years, others have refined Ilizarov's external fixator hardware and technique. Now many limbs with unhealed fractures, inadequate length, and angular or rotational deformities have been spared amputation thanks to that one patient who turned the wrench the wrong way. Anybody could do that, but it takes a genius to recognize the implications and appreciate that adversity is just opportunity in disguise.

Before shining light on the next orthopedic giant, Masaki Watanabe, some background information is in order. It is possible that soon after primitive humans started walking erect, their curiosity increased because they could see farther. Nosiness led them to peek into caves or to drop back onto all fours and peer down badger holes. Looking into their family's mouths and ears soon followed. Many generations later, their progeny developed metal tubes and glimpsed human interiors through all of our natural orifices. Lighting was an issue from the beginning, though, and the torch that satisfactorily illuminated the cave would not suffice in proctology clinics.

This changed in 1879 with Edison's invention of the incandescent light bulb. Just seven years later, two German doctors were lighting up bladders' interiors with tiny bulbs on the ends of steel tubes through which they squinted. Heat from the bulb and risk of breakage, however, posed problems. Nonetheless, enterprising doctors began poking holes in the skin and exploring the abdomen and chest with lighted tubes. In 1912, Severin Nordentoft, a Danish doctor, extended this concept to joints and coined the word *arthroscopy* (joint view). Multiple investigators around the world then refined and continue to refine the technique, especially for the problem-prone knee.

Prior to antibiotics, tuberculosis occupied much of an orthopedist's time. This was particularly so in Japan, where squatting and kneeling have long been cultural imperatives. In 1918, Kenji Takagi began using a bladder scope to examine tuberculous knees. His idea was to develop an early treatment that would preclude the awkward outcome of an entirely stiff knee. Over the next 20 years, he designed and tested 12 versions of arthroscopes that were progressively smaller in diameter and that incorporated better optical systems, but none of them were entirely practical.

After World War II, Takagi's student, Masaki Watanabe, took up the quest and continued to make design improvements. In 1957, Watanabe presented a film describing his work, first to an international orthopedic meeting in Spain and then to major European and North American orthopedic groups on his way home to Japan. The response was tepid at best.

Undaunted, Watanabe pressed on. His twenty-first version finally provided an adequate view and good focus even though it required that he grind each lens by hand. By 1958, this version became the world's first production arthroscope. But breakage of the incandescent bulb on the tube's tip continued to be problematic. Watanabe began receiving international visitors interested in learning his technique; but when they returned home, began using it, and reported their results, peer criticism, if not ridicule, prevailed.

In 1967, the twenty-second version for the first time incorporated a novel fiber optic cable. Now the hot, fragile light bulb could rest 10 feet away from the operative field and transmit "cold light" into the knee joint via thousands of bundled tiny glass strands.

Watanabe developed at least three more versions to further address the conflicting goals of better illumination and visualization versus smaller diameter scopes that could probe the deepest recesses of small joints. His final version was less than one-twelfth of an inch in diameter—about the diameter of a coat hanger wire. Later came miniaturized television cameras that attached to the arthroscope. A TV monitor displayed the images, allowing the residents, nurses, and students in the room to see what the surgeon saw. No longer did they have to stare at the back of the surgeon's head as he squinted into an eyepiece attached to a narrow straw. Patients when awake could watch, too, and a video recording of the event later allowed their families untold hours of viewing pleasure. Well, maybe minutes.

International interest began to grow along with further advances in arthroscopic instrument and scope designs. At first, every proce-

*Arthroscopes have undergone additional development since the pioneering work of Masaki Watanabe. This modern example is less than one-eighth of an inch in diameter and can even enter small joints in the hand and foot.* NanoScope.

dure was merely diagnostic, and the surgeon followed it immediately with an open exploration of the joint to directly see and treat whatever pathology the arthroscope had revealed. Tiny nippers and shavers, first manual and then also powered, began to allow for arthroscopic *treatment* as well as diagnosis. Current techniques and instruments even allow the surgeon to place and tie sutures inside a joint. Such minimally invasive surgery allows for faster and more successful rehabilitation. Because the knee joint is large, the innovations started there, but now orthopedists routinely apply these techniques to the shoulder, elbow, wrist, hip, and ankle joints. Undoubtedly, our caveman ancestors, torches and clubs in hand, would be pleased with what they started.

In the 1950s, about the same time that John Charnley was perfecting total joint replacements in England, American Paul Harrington addressed a vexing spine problem. For background, a snake naturally curves its spine repeatedly from side to side to slither along. By comparison, a human's spine is not as flexible. It can bend a little from left to right but is normally straight when its owner stands tall. If a human spine develops a curve to the side that does not go away when standing at attention, the bend is unbalanced and tends to progress and cause shortened stature, an unsightly humpback, and compression of the heart and lungs inside a twisted rib cage.

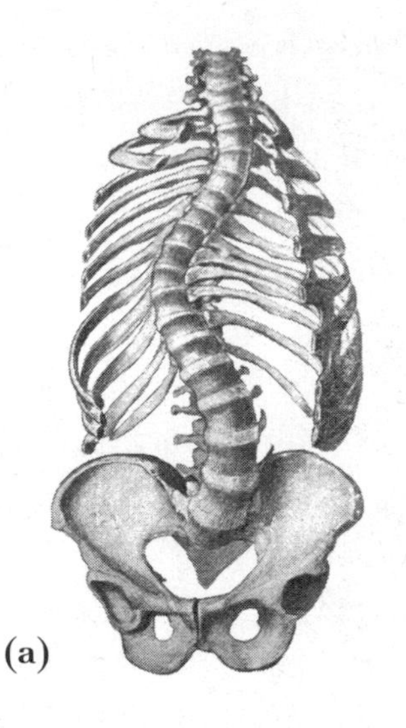


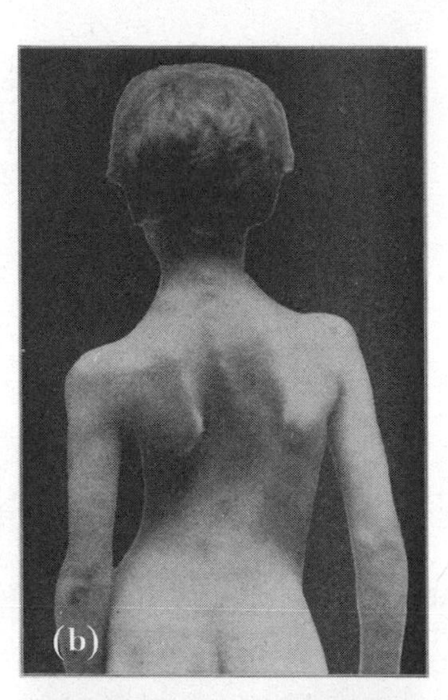


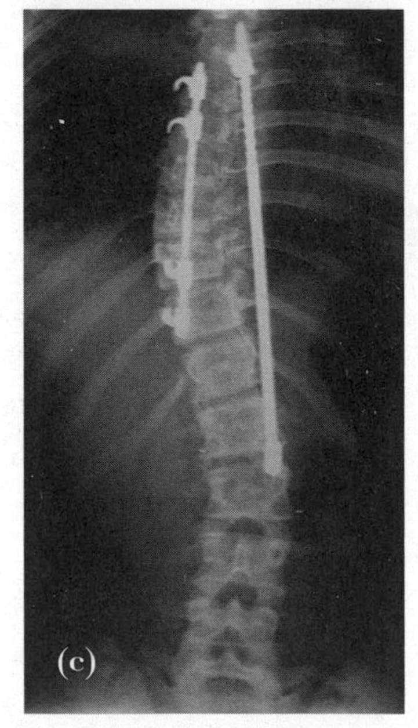


*(a, b) A side-to-side curvature of the spine is unsightly, twists the ribs, and stresses the heart and lungs. (c) Harrington's rods expand the curve on its concavity and compress the vertebrae together on the convexity to stabilize the spine in a functionally and cosmetically improved alignment.* (A) ROYAL WHITMAN, *A TREATISE ON ORTHOPEDIC SURGERY* (PHILADELPHIA: LEA BROTHERS, 1903); (B) JOHN RIDLON, HUGH OWEN THOMAS, AND ROBERT JONES, *LECTURES ON ORTHOPEDIC SURGERY* (PHILADELPHIA: E. STERN, 1899); (C) PAUL R. HARRINGTON ARCHIVES, UNIVERSITY OF KANSAS MEDICAL CENTER, KANSAS CITY, KS.

A slow death can ensue from compromised circulation and breathing. This side-to-side spinal curve is called scoliosis. In the mid-twentieth century, polio-induced muscle imbalance was the main cause of this debilitating deformity, and orthopedists tried various spine-stretching exercises and braces to correct or at least halt the deformity's progression. As you might imagine, maintaining alignment of the head over the pelvis with metal struts, leather straps, and horsehair pads proved to be impossible. Comfort was certainly not a consideration. Attempts at surgical correction were similarly tortuous.

Although not by choice, Paul Harrington happened to be at the right time in the right place to make a difference. He had grown up in Kansas and played on three consecutive championship basketball teams at the University of Kansas. Thereafter, he attended medical school and completed orthopedic surgery residency training in Kansas City. On return from World War II overseas duty, job opportunities were scarce. He found one in Houston that nobody else wanted—surgeon for the polio clinic.

Polio was then epidemic, its viral cause still unknown and the preventive Salk vaccine a decade away. Harrington, confronted with a large number of children and teenagers with post-polio scoliosis, joined forces with the clinic's brace makers. They made stainless steel hooks that Harrington surgically attached to the spine above and below the curved sections. He connected the hooks with a notched rod and ratcheted the spine straight, akin to jacking up a car. Harrington then fused the portion of the spine that was now spanned by the instrumentation.

After surgery, the patient remained immobile until the fusion solidified. This required months of bed rest followed by several more in a plaster cast from chin to hips. At times the hooks would dislodge, the rod would break, infection would occur, or the spine would not fuse. Undaunted, Harrington kept meticulous notes on each patient and gradually perfected the instrumentation, surgical technique, and postoperative regimen. This attention to detail eventually reduced the incidence of complications from 77 percent to zero over hundreds of patients.

In 1958, he presented his results to the annual meeting of the American Academy of Orthopaedic Surgeons, where his iconoclasm was met with astonishment, skepticism, and derision. A few orthopedists, however, chose to try his technique. Harrington insisted that they first visit him and observe the procedure. Acceptance was gradual. In 1960, *Time* magazine reported, "Some ailments seem almost preferable to their cures. A case in point is scoliosis, . . . treatment seems so punishing that [parents] cannot be persuaded to permit it even to save their children from permanent deformity. Last week Houston surgeon Paul Harrington, MD, was winning converts to a new and happier method."

As with most innovations, more advanced instrumentations have supplanted Harrington's system. These provide immediate stability and eliminate the need for postoperative bed rest and bracing. The new devices also allow for preservation of the natural front-to-back curves in the spine while correcting the dreaded side-to-side bowing. Although polio has all but disappeared in industrialized countries, Harrington's pioneering work still has purpose for spinal injuries and for scoliosis from other causes. Knowingly or unknowingly, Harrington went where he was needed, and his diligence paid off. He did not profit personally, because he never patented his bold and ingenious system, but society benefited greatly.

Charnley, Ilizarov, Watanabe, and Harrington advanced orthopedics through their sustained efforts to perfect devices in the machine shop and to improve surgical techniques in the operating room. Two other pioneers made their marks instead in research laboratories.

After completing military service in World War II and then his orthopedic surgery residency in Boston, Marshall Urist returned to his native Illinois and joined the faculty at the University of Chicago. There he partnered with a physiologist, and they focused their laboratory research on bone growth and bone grafting. Urist noted that new bone would form not only around a graft but also at times in muscle tissue some distance away. He surmised that some chemical messenger must be stimulating local cells to begin producing bone. Thereafter, he directed his research to isolate and identify that messenger. In the mid-1950s, Urist moved to Los Angeles and spent the remainder of his career at UCLA.

Urist's research assistants made regular visits to the slaughterhouse and returned with hundreds of pounds of beef bones. Under Urist's guidance they crushed and then processed the bone, first to remove the calcium and then to tease apart the proteins. Repeating this arduous process again and again, Urist was finally able to reduce a heap of bones to a tiny dot of bone-stimulating protein in the bottom of a test tube. When injected into tendon, brain, and fat, these chains of amino acids, known as growth factors, would induce the local cells to form bone.

Urist called this growth factor bone morphogenetic protein, and it is known worldwide today at BMP. Progress at characterizing and testing the effectiveness of BMP was slow, however, because it existed in minute quantities and took weeks to isolate and purify. A research fellow unwittingly discovered a shortcut. Until then, the isolation steps had been performed at room temperature. Wanting to leave for a weekend camping trip before a batch of bone could be completely processed, the assistant refrigerated the mix and took off. The next week, that batch yielded a far greater quantity of BMP than what previously could be harvested.

In due time, Urist, his former fellows, and others characterized the specific chemistry of BMP, which turned out to be a family of closely related growth factors, all of which stimulate bone formation. Investigators presently coerce bacteria to produce BMP, which is now commercially available and approved for certain clinical applications. It hastens healing of recalcitrant fractures and enhances complete fusion of spinal segments during treatment of low back and neck pain. For spinal fusions, surgeons now combine specially designed hardware that securely fixes one vertebra to the next with BMP-impregnated cadaver bone graft. These advances have markedly increased the success of this often risky procedure. I wonder where we would be today if the lab assistant had stayed home that weekend.

In the same mid-century era that the aforementioned orthopedic giants were innovating, Jacquelin Perry developed what other surgeons might consider a career-ending disability. Undaunted, she changed focus and ended up aiding untold numbers of individuals who limped; and in the course of doing so, she inspired many women to become either orthopedists or physical therapists.

At age 10, Perry knew that she wanted to be a doctor but took a circuitous route and studied physical education at UCLA. When World War II broke out, she joined the army, trained as a physical therapist, and worked in a rehabilitation hospital. This did not entirely suit her, because she felt that some of the treatment orders were incorrect, and she wanted to be able to make her own decisions. That itch led to medical school, with graduation in 1950 and subsequent residency training in orthopedic surgery. She was among the first 10 women to be board certified by the American Board of Orthopaedic Surgery.

Dr. Perry accepted a position at Rancho Los Amigos Rehabilitation Hospital in Southern California and continued working there until the week before her death at age 94, even though wheelchair bound in later years from Parkinsonism. When she started, polio was rampant, and children, weakened by the disease to the extent that their breathing was labored, were placed in iron lungs. They were certainly too frail to undergo any spinal stabilization surgery that might allow them to sit up and reduce their respiratory distress. Dr. Vernon Nickel and his new associate, Dr. Perry, developed a strange device to support these fragile necks by spanning them externally with bars that ran from a metal ring screwed into the skull to a vest worn snuggly on the chest. (As grisly as that may seem, the screws, placed under local anesthesia and not fully penetrating the skull, did not subsequently hurt.)

With the neck stabilized, Nickel and Perry then could operate safely and bridge the unstable spinal segments with bone graft. This advance was revolutionary, but Dr. Perry's surgical career was brought to an end by a neck problem of her own: she began experiencing extreme dizziness every time she turned her head.

Undaunted, she refocused on another highly prevalent problem at Rancho—limping, either from polio, cerebral palsy, stroke, or other neuromuscular disorders. For the remainder of her long career, she was a relentless problem solver, going back and forth repeatedly between the laboratory and the clinic. She studied gait, both normal and abnormal, wrote the definitive text on the subject, and became the maven of movement. Individuals with lifelong problems could now be approached both operatively and nonoperatively on a scientific basis. Also benefiting were individuals rehabilitating from joint injuries and athletes interested in improving their form. For example, Dr. Perry did not play golf but observed a physical therapist's swing and offered advice for improvement. It worked.

Dr. Perry's thirst for knowledge and passion for helping patients was infectious, and although coworkers and students variously described her as "no-nonsense" and "intense and stern," she exhibited "tough love" and had many positive, career-forming and defining influences on female orthopedic surgeons and physical therapists who came to Rancho to learn from her. Respectfully and appropriately, they titled her the "Grand Dame of Orthopedics."

✦✦✦✦

Thousands of other pioneers with various combinations of the curiosity, creativity, and tenacity manifested by Charnley, Ilizarov, Watanabe, Harrington, Urist, and Perry have also advanced the understanding of bone disease and treatment. One achievement stimulated the next. Some examples are in order.

*Chapter 7*

# ORTHOPEDIC INNOVATIONS

CONSIDER THROWING A BALL, HOLDING A pen, or fastening a button. With your thumb positioned to do these maneuvers, notice that its contact surface is opposite to and facing the contact surfaces of your other fingers. In this position, it opposes their free movement and allows for pitching strikes, signing checks, and getting dressed. Since the thumb can move in and out of this position, we call it opposable. Humans rarely give this amazing motion a conscious nod; but when combined with a large brain and brewed for several hundred thousand years, civilization ensued. Aided by opposable thumbs, it was only a matter of time before humans went from drawing on cave walls to sending Tweets, from hurling rocks at hungry lions to pitching against the Tigers, and from stitching hides to operating sewing machines.

Our opposable thumbs are ultimately responsible for everything human-made. Various languages celebrate its function. *Shast* in Farsi means both "sixty" and "thumb," signifying that it accounts for 60 percent of the hand's function. In Turkish, thumb is *bas parmak*—"chief finger." And in Latin it is *pollex*, which is derived from *pollere*—"to be strong." Isaac Newton marveled, "In the absence of any other proof, the thumb alone would convince me of God's existence."

Because this wonderful digit sticks out somewhat awkwardly and is involved in most manual activities, it is at high risk for injury. Those with severely disabled thumbs immediately appreciate the complexity of civilization and the difficulties of thumbless interaction. In case of the loss of any other single digit, by contrast, the three remaining can pretty well do the work of four.

Understanding the thumb's importance, hand surgeons work all night if necessary to piece an injured one back together. At times a completely amputated thumb can be successfully replanted. Although a replanted thumb is never normal in its sensibility, motion, and strength, it is certainly better than no thumb at all.

There are three major reconstruction techniques available when surgeons cannot repair or replant this critical part. I call them beg, borrow, and steal. For the first one, the surgeon and patient persuade (beg) the thumb stump to lengthen. The surgeon places steel pins transversely through the residual bone at both ends, attaches the pins to an expandable steel frame, and cuts the bone between the pins. Then by turning a little knob à la Ilizarov, the patient lengthens the frame in minute increments on an hourly basis over 6 to 8 weeks. The bone and the surrounding muscles, tendons, nerves, and skin hardly know they are being stretched. They just think there is some rapid growth underway, and they rally to keep up.

When the thumb is back to a functional length, the surgeon inserts a bone graft in the gap. The advantage of this lengthening procedure is that it avoids borrowing or stealing. The disadvantage is that the thumbnail and any missing joints remain absent.

The second technique capitalizes on the fact that the index, middle, ring, and small fingers work similarly, and losing one of them is not nearly as debilitating as losing a thumb. As a result, what can be considered when the indispensable thumb is missing and there are four underemployed friends stationed nearby? Borrow one! Because of its proximity, the index finger usually gets the call. The surgeon shortens and rotates it into the thumb position. This procedure is more complicated than remnant lengthening and leaves a four-digit hand. The advantages of borrowing for thumb reconstruction are that the new thumb has a nail and the convalescence is short.

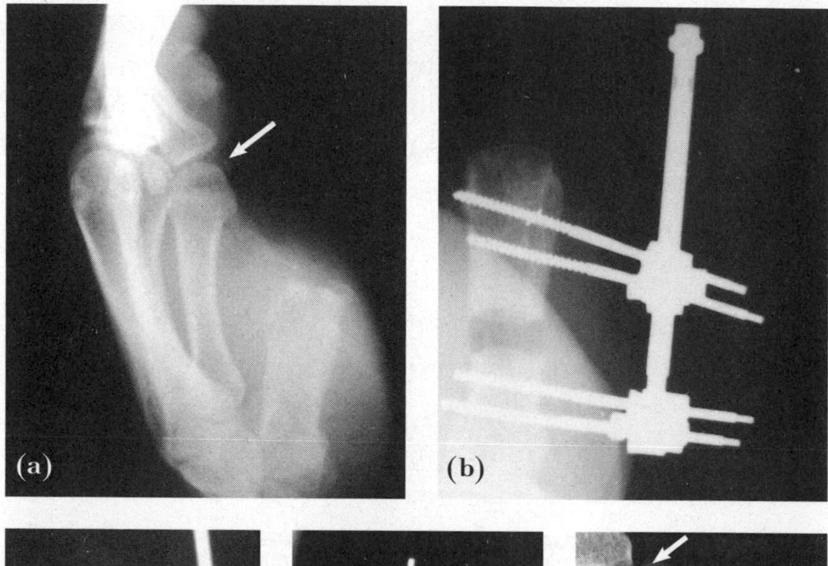


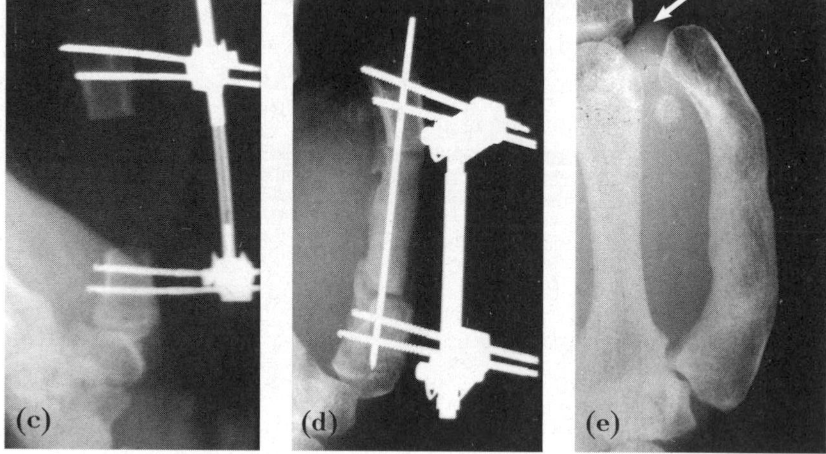


*This thumb amputation stump underwent gradual lengthening to restore sufficient length for performing pinch and grasp activities. (a) Before treatment, the thumb remnant is markedly short, as noted by comparing it with the level of the index finger's first joint (arrow). (b) At the first operation, the distraction device is placed, the bone is cut across, and the ends are gently stretched apart. (c) Over 6 to 8 weeks, incrementally elongating the threaded bar on the distraction device draws the bone farther apart and brings the skin, tendons, nerves, and blood vessels with it. (d) At the second operation, a bone graft from the hip fills the gap between the ends of the lengthened bone. (e) Once the bone graft has healed, the hardware is removed and functional activities can begin. The arrow indicates the length gained compared with before treatment.*

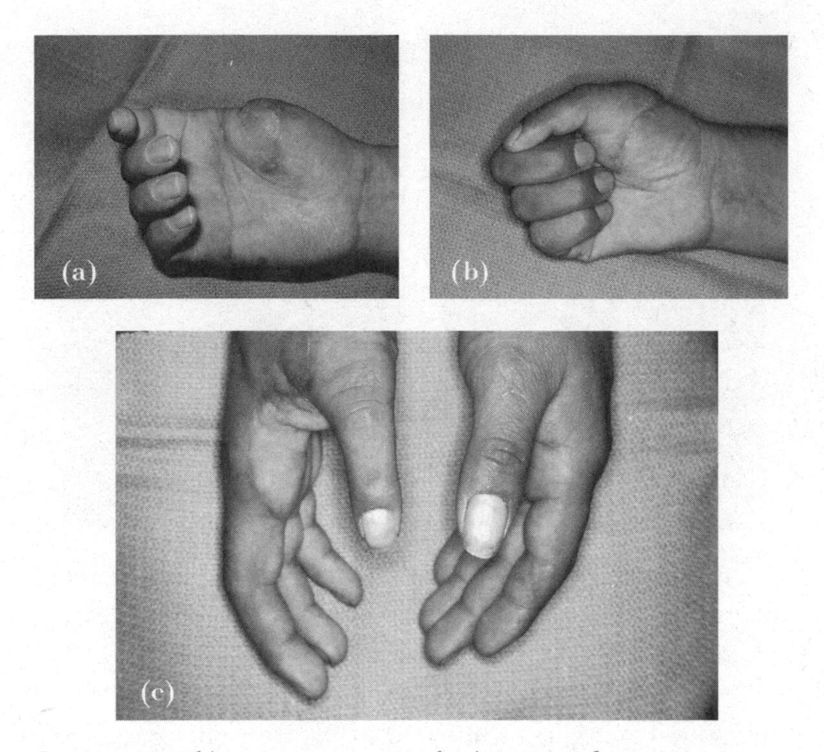


*Preoperative (a) and postoperative (b, c) images of a thumb reconstruction performed by shortening and rotating the index finger into the missing thumb's position.*

At times however, multiple digits may be missing from the same hand or the patient has a need for a five-digit hand. Either way, there is no thumb remnant to beg nor finger to borrow. The surgeon can resort to theft. The victim is the big toe. Its shape is almost identical to the thumb's, so it is the favored mark, although its absence leaves an unsightly crime scene. Stealing the second toe makes for a rather scrawny thumb but leaves a pleasantly contoured foot, so this is usually chosen in cultures where people remove their shoes indoors. Absence of either the first or second toe has minimal effect on the owner's ability to walk and run.

Toe-to-thumb transfer is a five-to-ten-hour operation requiring honed microsurgical skills. In the foot, the toe's nerves, arteries, veins, and tendons are identified, isolated, and cut along with the bone. The meticulously amputated toe is then attached to the hand

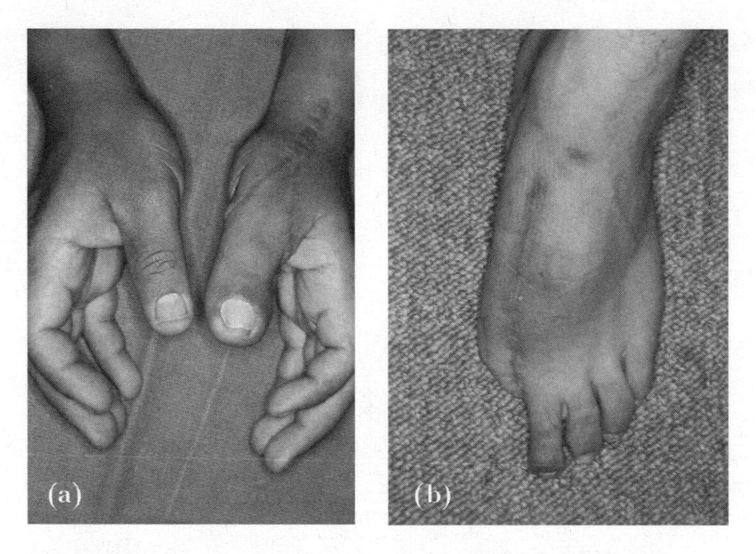


*This patient underwent a big-toe transfer for reconstruction of his traumatically amputated left thumb. The foot remains entirely functional even in the absence of the big toe.*

by attaching like tissue to like tissue. The surgeon secures the cut surface of the toe bone to the remnant of the thumb skeleton, usually with steel pins. The blood vessels are one-sixteenth to one-eighth of an inch in diameter and are precisely sutured using nearly invisible suture material. The carefully stitched vessels allow blood to flow through but not leak out. With circulation restored, the bone ends blithely heal together, unaware that there has been a comingling of foot and hand parts. Over months, new nerve fibers grow into the transferred toe to provide sensation. Some call the combo a thoe.

✦✦✦✦

REMEMBER THAT CHILDREN were the essence of orthopedics as it got its start in the eighteenth century. Doctors focused treatments, all nonoperative, on the skeletal effects of tuberculosis, polio, and rickets. As these diseases were eliminated and as surgery became safer thanks to general anesthesia and sterile technique, more complicated treatments were possible. One involves application of Ilizarov's external fixators to make little people taller.

Italian orthopedic surgeons introduced the Ilizarov technique to the Western world for fracture management and soon began using it to make extremely short people taller. Yes, some consider this treatment merely cosmetic, but others respond that the lengthenings are integral to the overall well-being of markedly height-challenged individuals. Early on, the operative technique required application of external fixators to both legs, cutting the shinbones in midshaft, and then gradually lengthening each side 2 to 3 inches over several months. The frames remained in place for additional months while the lengthened bones became sufficiently strong to withstand body weight. As a second procedure, some patients now choose to have their thighbones similarly lengthened. The good news: they may reach about 5 feet in height. The bad news: their arms then look disproportionately short. Accordingly, some individuals go on to have their upper arm bones similarly lengthened. Skin infections can track along the pins into the bone. Careful cleansing of the sites where the pins penetrate the skin is therefore obligatory, as is early and aggressive treatment of any pin site infection. Otherwise the procedure may swap one difficult problem for another.

In recent years, surgeons have shortened the time required for the Ilizarov frame by stabilizing the lengthened and still weak bones with rods placed in the bones' hollow interiors. Less time in the fixators not only is convenient but also reduces the risk of pin-track infections. An advanced technique for bone lengthening avoids use of the external frame altogether. Instead the orthopedist inserts a self-lengthening rod. By periodically holding a magnet against the skin over the rod, a slight rotation of the leg causes the rod to lengthen a bit, pushing the cut ends of the bone apart.

As you might imagine, bone lengthening for short stature is fraught with complications, and its application is reserved for rather stoic individuals with fully supportive families. Stretch too slowly, and the bone heals near its original length. Lengthen too quickly, and the bone gap does not fill in. Also the muscles, arteries, nerves, and skin need time to respond to the additional bone length, so intensive physical therapy is mandatory to maintain joint flexibility and pliability of the adjacent soft tissues. The treatment is not for the fainthearted patient, family, or orthopedist.

The next procedure also may not appeal to the squeamish. It is principally reserved for children with malignant bone tumors near the knee; and although it is rarely performed, it provides an example of how orthopedists can make lemonade out of lemons.

Skeletal cancers usually occur in the bones that are rapidly growing because that is where cell divisions are most numerous and where a mistake is most likely to occur. For teenagers, that means the bones that meet at the knee. Traditionally, the only effective treatment for a bone cancer in this area was an amputation close to the hip. This left the teenager with a short stump of thigh and a self-image-destroying limp because the absence of both the knee and ankle joints makes the use of an artificial limb clumsy and tiring even just for walking slowly. Especially for a teenager, such a stigma is likely to be physically and emotionally devastating.

An out-of-the-box solution begins with removing the cancerous segment of bone along with the knee joint but leaving the surrounding muscles, tendons, nerves, vessels, and skin intact. Stop here and the leg and foot would flop around, leaving the limb as dysfunctional as one with an above-knee amputation. Sure, the gap in the skeleton could be closed by merely attaching the ends of the remaining bones to one another and getting them to heal. That way, lower limb stability would be restored, but the limb would be knee-less and extremely short. Furthermore, the toes would peek out beneath the hem of a skirt or poke forward inside a pant leg—not something a teenager would easily accept.

Now comes a hat trick. Before closing the gap and fastening the ends of the remaining bones together, the pediatric orthopedist rotates the foot and ankle 180 degrees so that the heel points forward and the toes point backward. The skin, muscles, nerves, and blood vessels can tolerate this half turn, just as they do when a hand surgeon shortens and moves an index finger into the thumb position. After twisting the lower limb, the ankle joint then serves as a knee joint, and the foot is slipped into an artificial lower leg and foot. "Grotesque" and "bizarre" are common responses when people first hear about this technique; but the new knee bends in the correct direction, and the toes, pointing down inside the prosthesis, are not visibly obtrusive. Now with one normal knee and one ankle-knee, the owner can run and skate with little limitation. Go to YouTube and enter *Van Nes rotationplasty*. The video clips clarify the concept and demonstrate the result in a way that words fail.

✦✦✦✦

IN PAST CENTURIES, adults often did not live long enough to develop cancers. Rather they died young from tuberculosis, cholera, plague, typhoid, influenza, and similar infectious scourges, which are now usually muted or nonexistent in industrialized countries. And historically, when a bone cancer did occur, it may have already spread to vital organs and killed the individual before the primary source was recognized or before any treatment could be rendered. Even if recognized early, the only treatment was a swift limb amputation followed by the use of crutches or a wheelchair for life.

Even though the outcomes are not nearly so dire these days, bone cancers still develop in one of two ways. They can arise de novo in the bone, especially near teenagers' knees. These are "primary" bone malignancies. Most of them result from cell division gone awry in an osteoblast or osteoclast or in one of the blood-forming cells that reside in the bone's spongy interior. "Secondary" bone cancers arise from some other tissue in the body and spread to the bone, usually via the bloodstream.

If a bone cancer is identified early, limb salvage is frequently possible. This often calls for a multidisciplinary approach using some combination of surgery, radiation therapy, and chemotherapy. Limb salvage can preserve function far better than what could be expected from amputation and prosthetic fitting. One such salvage procedure is the knee excision and foot rotation technique just described. Its use, however, is limited to tumors near the knee and just for children and adolescents, because older blood vessels might not survive the obligatory 180-degree twist. What other salvages are possible?

Remember John Charnley and his pioneering work with total hip replacements? Orthopedic oncologists first extended his work by replacing entire thighbones that were cancerous. Now they replace other bones completely as well. These implants are custom-made to match the length of the removed bone, including the joint-shaped surfaces on both ends. In instances where the removed portion includes the knee joint, the implant can include a substitute hinge.

Problems with these metallic bones occur frequently and include infection, loosening, and metal fatigue with breakage, so such massive bone and joint replacements are typically reserved for low-demand patients. For children and adolescents with bone cancers near the knee, an excision, shortening, and rotation would likely be more durable. Nonetheless, some children with bone cancer do receive skeletal prostheses, but these will need to be replaced, perhaps several times, to maintain equal leg length as the unaffected side grows normally. On the horizon may be skeletal prostheses that "grow" in pace with the patient. These involve ingenious mechanisms inside the artificial bone that will allow lengthening with just a minimal operation at most.

As exemplified by artificial heart valves, ocular lenses, arteries, and joints, biomedical engineers strive to develop implants that are compatible with living structures. Biology, however, is an age-old judge and will eventually expose the weaknesses of even the best space-age body-part replacements. Even though skeletal implants certainly have their place and have spared many limbs with bone cancer that would have otherwise been amputated, they are not as good as the ones made from real bone.

The treatments for skeletal cancers described so far involve extensive surgery, but on just one bone. What happens when the disease has spread—for instance, when cancerous cells have already affected multiple bones or when they have taken over the blood-forming cells residing in the bone marrow?

One consideration focuses on strontium, a well-known industrial additive to ceramic glazes, magnets, and fireworks, where it creates deep red colors. Some toothpastes contain strontium, and it is also used in Europe to treat osteoporosis. It works in teeth and bones because it has chemical similarities to calcium.

There are four naturally occurring forms of strontium. Consider them brothers. They are all calm, well-behaved community stalwarts and differ from one another only slightly. Bone cannot tell the difference between calcium and strontium, so it readily incorporates either element when it needs building material. Once strontium becomes part of bone, however, it is permanent, whereas calcium comes and goes, depending on the heart's needs. The trace amounts of the four naturally occurring forms of strontium in our bones cause no harm, but they have at least 16 rather nasty, unstable stepbrothers. They are all radioactive and capable of mischief. That's both bad and good news. One of the stepbrothers, strontium 90, is a product of nuclear explosions. It is present in nuclear fallout and is unfortunately long-lived. Strontium 90 has performed only half of its radioactive mischief in its first 29 years of life, a quarter more in the next 29, and an eighth more in the next 29. After the Chernobyl disaster, strontium 90 fallout settled on pastureland as far away as Sweden and Scotland. Cows that ate the dusted grass carried the strontium 90 into their milk, which then became a permanent part of milk drinkers' bones. The radioactivity can cause bone cancer decades later. That, of course, is bad news, the magnitude of which is yet to be seen.

Conversely, the mischief of another stepbrother, strontium 89, can work to the body's advantage. This unstable relative is also radioactive but not nearly so long-lived. It performs half of its mischief in just 7 weeks, so more than 99 percent of its nastiness is gone after a year. Radiation oncologists treat some bone cancers with strontium 89. It settles in rapidly growing bone cells (that is, the cancerous ones) and kills them. That is great news—targeted radiation therapy.

The bone cancers I have described so far are diseases of the bone cells themselves, either osteoblasts or osteoclasts. There are also cancers of the blood-forming cells that reside in the bone marrow—leukemia, for example. These types of cancers cannot be cut out. Rather, doctors use combinations of chemotherapy and radiation therapy to kill all the marrow cells. Then through a vein, they either transfuse back the patients' own healthy cells (collected and preserved before cancer treatment) or transfuse healthy cells from another person—a bone marrow transplant. Either way, the infused cells find their way into areas of spongy bone and resume producing blood cells, minus the cancerous ones.

◆◆◆◆

IN THE MID- TO LATE 1800s, a remarkable confluence of advances changed everything regarding the treatment of fractures, and these innovations are also noteworthy. The discovery and application of general anesthesia allowed for more detailed and meticulous surgery. No longer was the fastest surgeon necessarily the best. New knowledge of bacteria and means of reducing the high incidence of operating room infections made surgery no longer life-threatening. X-rays for the first time allowed doctors to see the exact configuration of fractures and dislocations and plan treatment accordingly.

Even with these advances, however, early attempts to directly fix fracture fragments together with wire sutures or with metal plates and screws failed. The hardware frequently came directly from woodworking or metalworking shops, and no thought was given to its incompatibility with the gallons of salt water contained inside our waterproof skin. Either the metal corroded before the fracture healed or the body reacted vehemently to the metal or both. After observing this, investigators tried pegs, plates, and screws made of ivory, which the body tolerated better, but the implants were too brittle to be practical, which was fortunate for the elephants.

Stainless steel proved to be the answer. Metallurgists in the first several decades of the twentieth century perfected it, and orthopedists were quick to adopt this strong and nonreactive metal for their purposes. They began slipping rods through the hollow central portions of bones to support fractures internally. These rods ranged from the size of a matchstick, which were useful for stabilizing hand and foot fractures, to ones as thick as a pencil and half again as long to stabilize fractures in much longer bones.

Around the same time, Gerhard Küntscher, in Germany, devised a "nail" for internally fixing thighbone fractures. This implant was the shape of a three-leaf clover on cross section. This gave it the necessary rigidity to securely hold the fracture while allowing sufficient flexibility for it to find its way down the slightly curved central channel as Küntscher hammered it home. His work was only discovered in America when captured US pilots, whose broken thighbones had been fixed in Germany, returned home displaying Küntscher nails on their X-rays. It was a game changer. Previously, an unfortunate soul with a thighbone fracture had a pin drilled through the bone near the knee. The pin, sticking out through the skin on both sides, was attached via a rope and pulley system to weights that dangled off the end of the bed. This traction stabilized the bone ends against one another but necessitated 6 weeks of recumbency in a hospital bed. By then the fracture was sticky enough that the traction could be released and replaced with a cast, which spanned from armpits to toes on the injured side. Then the patient could be up on crutches until the cast was removed after another 6 weeks.

An even greater interest has surrounded the evolution of plates and screws applied to the surfaces—rather than into the central canals—of broken bones. Innovators have capitalized on the growing understanding of bone-healing biology, including the benefits of immobilizing the fracture site while at the same time preserving a robust local blood supply. The ingenuity displayed in modern hardware is amazing. The first successful plates and screws were stainless steel; now some are titanium, which like stainless steel is corrosion resistant. Titanium has another property that makes it useful for plating bone. If a plate is soft and easily bendable (think stick of gum), it does not stabilize the fracture whatsoever. Conversely, if a plate is entirely rigid (think knife blade) and holds the fracture absolutely still, the hydroxyapatite crystals at the fracture site are never mechanically deformed. Therefore, these calcium crystals do not generate any piezoelectric forces. And without that guidance, the cutting cones are unable to remodel and strengthen the bone. In this situation, the plate bears the load, which is fine unless the plate is later removed. Then the furloughed cutting cones may not get the back-to-work message soon enough. The bone is weak. It breaks again. The parlance for overprotecting the cutting cones is stress shielding, and less stress shielding occurs in bone with titanium than with stainless steel fixation because the titanium has a bit more give to it. By one means or another, the goal is to stabilize bone ends just enough to maintain fracture alignment while allowing some microscopic motion, enough to keep the cutting cones on the job.

Not only were all plates initially stainless steel, they were all straight (easy to manufacture) and thick (strong but bulky). These features were fine as long as the targeted bone was also straight. Furthermore, there had to be adequate muscle and skin to close the incision following placement of the bulky plate—not always easy. Plates now come in a vast array of shapes, curves, thicknesses, and widths—precontoured for specific applications. Plates with S-shaped curves and slight twists come in several sizes to fit right collarbones perfectly. Then there is a mirror-image set for left collarbones. These plates, however, are useless for fractures elsewhere. If you just glanced at these plates, you would think that they had spent an hour in a garbage disposal and had become randomly bent.

Screw technology has advanced along with plates (although all screws are still straight). Length, core diameter, thread diameter, thread pitch, and drive type (for example, Phillips, hex, star) all

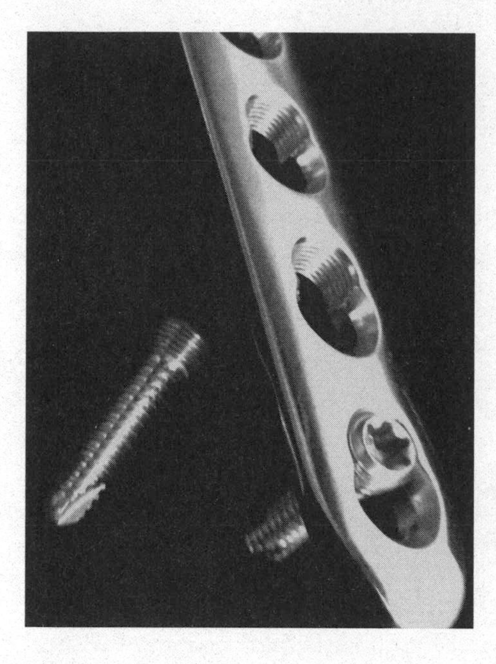

*This fracture fixation plate can accept either conventionally designed screws placed into the larger, smooth-edged portion of the holes or specialized screws that are threaded on their heads as well as on their shafts. Use of head-threaded screws allows for fixation of complex fractures that previously defied secure stabilization using conventional hardware.*

come into play. Some of the screws are headless, threaded along their entire length and designed to be entirely buried in bone. Others have heads that swivel on the shaft. Six to 10 screws of every configuration rest patiently in sterile trays stacked on the instrument table. An experienced scrub tech knows where each one is and delivers the specified one to the surgeon after confirming its length, diameter, and thread configuration.

Another interesting technology is each hole in the plate. These seemingly simple voids have undergone as much scrutiny and design as the plates and screws themselves. A recent innovation is locking screw technology, which allows the screw's head to thread into the plate as the screw's shaft threads into the bone. This rigid construct makes it nearly impossible for any movement to occur between any of the players: bone, plate, screws. As a consequence, smaller plates and fewer screws are required even in shattered osteoporotic bone. Locking screw technology accounts for the fact that wrist fractures, traditionally treated in casts for 6 weeks, are now often treated with plates and locking screws. No cast is necessary, even in frail oldsters with fragile bone.

◆◆◆◆

THE INNOVATIONS DESCRIBED, from shortening and rotating bones to implanting metallic supports, all require precise awareness of the affected bone's shape and alignment. In addition, once a treatment is applied, the progression of healing requires monitoring; but bone is concealed, and we are hesitant to expose it and look directly. What are suitable compromises?

*Chapter 8*

# PICTURING BONE

EARLY ANATOMISTS AND PHYSICIANS TOOK little interest in bones, so bones were rarely and only crudely depicted. Several reasons accounted for this indifference. From Galen's time in about AD 150 until the Renaissance 1,500 years later, the Grecian notion prevailed that reason superseded observation, so why bother looking at the real deal or even a drawing? Furthermore, in the Middle Ages, elective surgery amounted to little more than bloodletting, so there was no need to understand anatomy. With rare exceptions, the Church also forbade human dissection, so there was little *opportunity* to understand anatomy. Hence, medieval depictions of human anatomy were sketchy and often based on imagination or on dissections of bears and monkeys.

Because reasoning superseded observation for hundreds of years, medieval anatomists reconciled differences between their findings and Galen's writings in one of two ways. They either discounted their own observations and sided with Galen or claimed that the anatomy had changed since Galen's time. As an example, Galen wrote that the thighbone was curved, although he may have been looking at a bear's. Anatomists later noted that the human thighbone was straight. Out of respect to Galen, they discounted their own observations and rationalized that the straightening might have occurred over the intervening centuries since Galen owing to the habit of wearing "cylindrical nether garments."

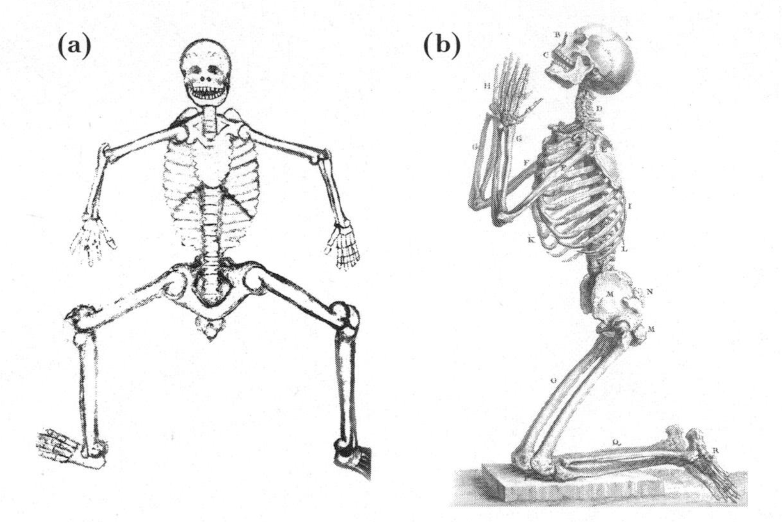


*(a) An unknown anatomist drew this image in 1323, long before human dissection was commonly performed. The breastbone is highly stylized, and the collarbones and shoulder blades are depicted erroneously as a solid ring, as is the pelvis. (b) William Cheselden, 410 years later, quite accurately detailed the precise skeletal anatomy. He posed the skeleton in prayer to fit the largest possible image on a single page of his atlas,* Osteographia. (A) J. G. DeLint, *Atlas of the History of Medicine* (New York: Hoeber, 1926), p. 27.

Fortunately, the invention of the printing press helped bring the Dark Ages to an end. The first known human anatomical print appeared in 1493. Over the next hundred years, learning in Europe flourished and included the establishment of both observational science and the first medical schools. Human dissection, typically using the bodies of executed criminals, became an infrequent yet routine part of medical school curriculum. At the same time, Leonardo da Vinci, Michelangelo, and others grasped for the first time the concepts of perspective and shading. Leonardo noted to himself, "Make a demonstration of these ribs in which the thorax is shown from within, and also another which has the thorax raised and which permits the dorsal spine to be seen from the internal aspect. Cause these two shoulder blades to be seen from above, from below, from the front, from behind, and forward." Such attention to detail led to the development of printed atlases that accurately represented human anatomy and thereby made anatomical knowledge widely accessible. By 1700, bones had become the visual icon for all of human anatomy. One anatomist in the 1700s deserves special attention because he focused solely on bones.

At age 15, William Cheselden apprenticed himself to a noted London surgeon. Seven years later, in 1709, he emerged as a surgeon himself. Unable to immediately develop a practice, he taught anatomy and eventually turned his class notes into a book, *The Anatomy of the Human Body*. It was wildly successful, partially because it was in English rather than Latin, which was the norm at the time. The book proceeded through 13 editions and was the go-to source for surgical anatomy for 100 years.

Based on his deep understanding of anatomy, Cheselden became an adept surgeon—setting fractures, removing cataracts, and especially extracting bladder stones. He devised a new approach to the bladder and could remove a stone in less than a minute after the initial incision. Since general anesthesia had not yet been discovered, lightning-fast surgery not only shortened the patient's agony but also reduced the procedure's mortality rate to a previously unbelievable 10 percent. Cheselden's reputation spread, and he became the preeminent surgeon in England if not the world.

Further successes ensued. Queen Caroline appointed Cheselden as her personal surgeon. Cheselden was instrumental in persuading King George II to dissolve the 200-year-old charter that had formed the Company of Barber-Surgeons. Cheselden's circle of friends included Alexander Pope and Isaac Newton.

He is best remembered, however, by what might be considered a failure. Understanding that skill in surgery required a thorough understanding of anatomy, in 1733 he published *Osteographia, or the Anatomy of the Bones*. This was the first and only volume of what he anticipated to be a three-volume set of anatomical illustrations. *Osteo-*

*graphia* took several years and £17,000 to complete. This was the first book devoted solely to bone anatomy. It sold only 97 copies. Yet *Osteographia* is a treasure of anatomy, artistry, and human culture, which is why I respectfully include some of Cheselden's plates in this book.

In the Middle Ages, anatomical drawings had been symbolic and crude. During the Renaissance, however, artists began to understand perspective and shading—techniques that were critical to rendering three-dimensional objects accurately onto paper. Even so, just a slight shift of the artist's head or an urge to highlight a shadowed surface could distort the result. Cheselden wanted to avoid this outcome. Surgeons needed absolutely accurate representations of the skeleton with perfect subtleties of each contour. Medical photography would not be available for another hundred years; until then, dissections and drawings were the only ways to learn and teach anatomy.

Under Cheselden's guidance, two artists accomplished his aim of drawing with absolute accuracy. Unique at the time for medical illustration, they did so by suspending a bone from a tripod, which they placed in front of a large wooden box. The box had a tiny hole in one end where light and the image of the bone could enter. The artist sat with his head and arms in the other end of the box and traced the detailed features of the bone image on glass plates, which Cheselden then converted to engravings. Cheselden's generation knew this box device as a camera obscura. Perhaps it is better known today as a pinhole camera—a giant one.

*Osteographia* is one of the all-time great anatomy atlases in scope and elegance. The camera obscura, depicted on the book's title page, signals the work's accuracy. The etchings' exquisite detail, their arrangement on each page, and the lack of overlying labels or lines convey refinement and sensitivity. Cheselden posed a complete human skeleton on its knees with its hands in prayer. He so positioned this figure to preserve the bones' relative sizes while economically fitting the largest possible image onto one page. Cutaway images illustrate the interior structure of bone and the means by which bones grow longer, as seen on p. 11.

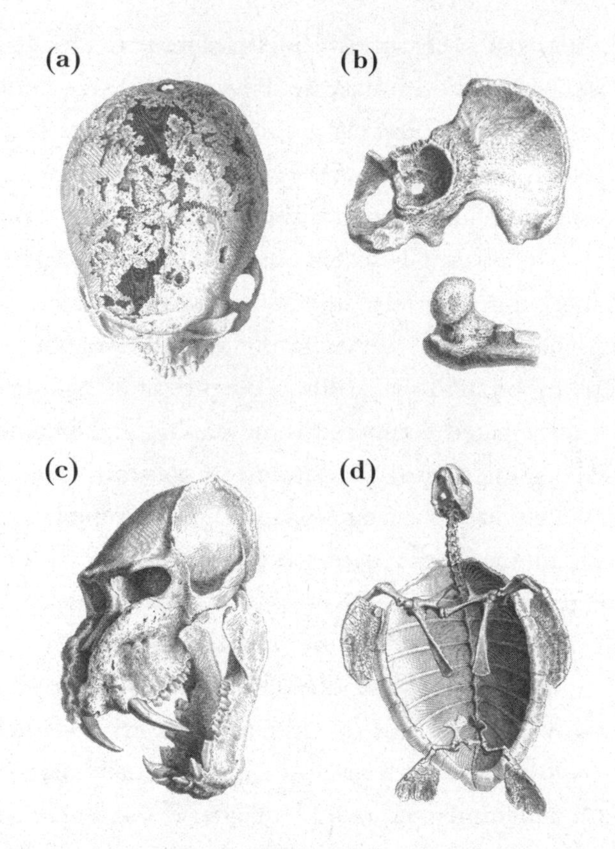


*In addition to depicting normal human bone anatomy in his 1733 atlas, William Cheselden included images of common bone disorders. (a) Syphilis has devastated the skull. (b) An abscess of the hip joint has eroded an opening through the socket and has destroyed the joint surfaces. Cheselden also depicted the skeletons of other animals to demonstrate similarities and differences between species. (c) Tiger. (d) Sea turtle.* WILLIAM CHESELDEN, *OSTEOGRAPHIA, OR THE ANATOMY OF THE BONES* (LONDON: W. BOWYER, 1733).

Cheselden interspersed depictions of normal anatomy with images of diseased bone. Plates illustrate a leg bone fracture healed with deformity (see p. 51), a skull riddled from syphilis, a chronically infected arm bone following a gunshot injury, and an arthritic hip joint. All would be instructive to doctors. The text is sparse. Cheselden knew that the images could tell their own story.

For comparison, *Osteographia* includes several images of other animals' skeletons. The animals are naturally posed—a dog and cat snarl at one another, a bear claws the bark off a tree. It is obvious that Cheselden loved bones. *Osteographia* is his enduring legacy. Nonetheless, medical illustration was eventually to be eclipsed by photography. In a way, Cheselden anticipated the change—he used a camera (obscura) to ensure the accuracy of his drawings.

Just as Cheselden, in Britain, anticipated photography by precisely detailing bone in the 1700s, a century later Nikolai Pirogov, in Russia, anticipated computed tomographic and magnetic resonance imaging for capturing anatomy in planes that cut through the body. A fast learner, Pirogov graduated from medical school at age 18 and enjoyed a stellar career as an innovative surgeon. He was also an anatomist and in the 1850s published his four-volume *Topographical Anatomy of the Frozen Human Body*. This atlas uniquely included multiple carefully rendered cross-sectional drawings of the body. Pirogov achieved this by freezing cadavers and then sawing them through at various levels and in different planes—precisely what computed tomographic and magnetic resonance images do electronically today.

The advent of photography, however, cut short the utility of Cheselden's and Pirogov's innovations for anatomical illustration. Just as many innovators contributed piecemeal to orthopedic advances, the same is true for photography, which had its beginnings in the 1820s. The first medical photograph, which was of a woman with a huge goiter, was taken in 1847. Through the remainder of the nineteenth century, photographs accurately recorded thousands of medical conditions.

Color photography followed, and specialized lenses captured close-up images of surface lesions and shots from inside ears, eyes, throats, colons, and even joints. Taking photographs through microscopes became routine. Photography of bones in their living state, however, remained elusive. This changed on November 8, 1895, the day in Würzburg, Germany, when Wilhelm Conrad Röntgen discovered X-rays and unknowingly vaulted orthopedics into the modern age.

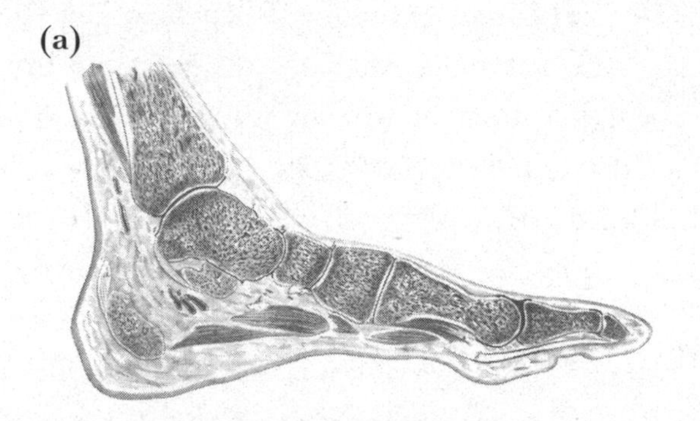


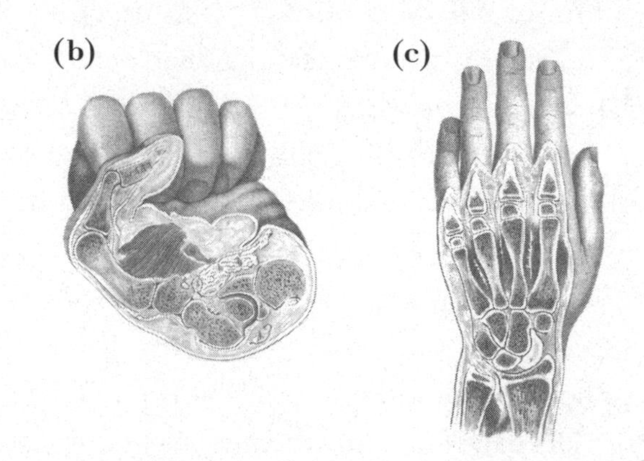


*Preceding computed tomographic and magnetic resonance imaging by over 100 years, in the 1850s Nikolai Pirogov produced an atlas of cross-sectional anatomy to facilitate conceptualizing the anatomy three-dimensionally.* Nikolai Pirogov, *An Illustrated Topographic Anatomy of Saw Cuts Made in Three Dimensions across the Frozen Human Body (Atlas, Part 4)* (St. Petersburg: Typis Jacobi Trey, 1852–1859).

Around that time, Thomas Edison, Nicola Tesla, Röntgen, and others were experimenting with glass vacuum tubes to determine the effects of passing electricity from one plate inside the tube to another. To exclude any interference of light coming from the tube, Röntgen sealed one in cardboard, darkened the room, and turned on an electric generator. To his great surprise, a nearby piece of cardboard that he had previously coated with a photosensitive chemical began glowing. When Röntgen shut the current off, the glow disappeared. With the weekend upon him, he repeated the experiment in various ways and made preliminary notes. Over the following weeks, he ate and slept in his laboratory and studied this unknown ray, which he labeled "X," the mathematical symbol for an unknown. He learned that X-rays passed through books, no matter how thick and that coins cast a shadow on the photosensitive board as did the bones of his own hand. Six weeks later, Röntgen shared the secret with his wife, who allowed him to take a 15-minute exposure of her hand, the first orthopedic X-ray. When she saw the image of her hand skeleton, she exclaimed, "I have seen my death." A week later, Röntgen presented his findings, including the image of his wife's hand, in a paper titled "On a New Kind of Rays." This caught the immediate attention of physicists, who alerted the lay press. The discovery made the front-page headline news within a week of Röntgen's public announcement.

Because cathode-ray tubes were well known and easy to make, many investigators contributed to the understanding and practical applications of X-rays. Interest was intense and advances were rapid. Less than three months after Röntgen's public announcement, an enterprising electrical contractor and avid photographer opened a laboratory offering diagnostic services. A month later the initial issue of the first radiological journal, which would become the *British Journal of Radiology*, was published. You name it, somebody was taking X-rays of it. By 1898, an atlas had appeared that depicted the growth plates in the hand and wrist, which orthopedists and radiologists still use to determine a child's or teenager's bone age.

Perhaps with equal parts of chagrin and enlightenment, doctors discovered that many injuries that they had always called dislocations were actually fractures and that manipulations directed at realigning skeletal injuries were often ineffective.

Röntgen received the Nobel Prize in 1901, the first one ever awarded for physics. Röntgen not only gave the reward money to

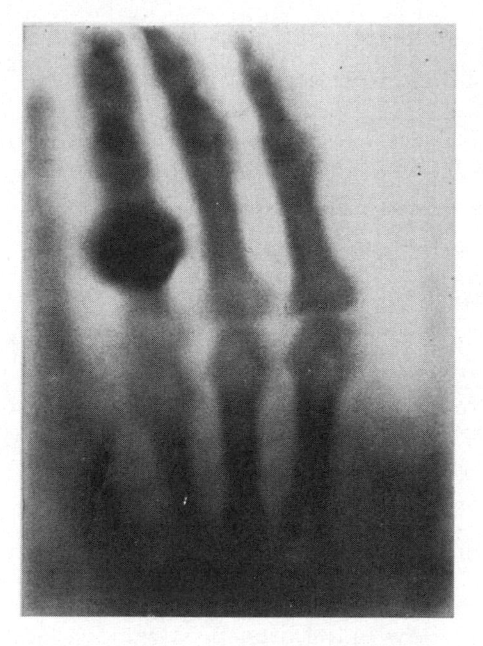

*Wilhelm Conrad Röntgen took this X-ray of his wife's left hand, including her wedding ring, in 1895. It is the first permanent radiographic image ever taken of a human.* NATIONAL LIBRARY OF MEDICINE.

his university but also refused to take out patents on his discovery to allow for widespread application.

I suppose when X-rays were in their infancy, patients were asking, "Now that you have finished obtaining a thorough medical history and performing a careful physical examination and telling me that you know with assurance what is wrong, aren't you going to order an X-ray, Doctor?" This question implied a lack of trust in the doctor's diagnosis unless he threw in a high-tech, trendy X-ray evaluation. Over the twentieth century, doctors and patients came to understand pretty well when an X-ray study could help make the diagnosis or plan treatment and when one would be superfluous. For instance, it is routine thinking today that a sore tooth most likely deserves an X-ray while a sore throat probably does not. In general, X-rays reveal structures that contain a high level of calcium—enough calcium to cast a shadow in the X-ray beam. Bones, teeth, hardened arteries, and kidney stones come to mind.

Doctors have learned to order X-rays with some caution because radiation damages living tissues. That fact required discovery, and the damaging effects of early X-ray evaluations were slow to reveal themselves. Since X-rays could not be seen or felt, investigators naturally considered them harmless. Both Tesla and Edison experimented with X-rays and observed that their eyes became irritated, but neither drew a correlation between the radiation and their symptoms.

For convenience, dentists originally held the film with their fingers when shooting X-rays of teeth. It was decades later before the skin on their hands dried, cracked, and became cancerous. I remember in the 1950s enjoying watching my toe bones wiggle in the fluoroscopy unit in the shoe department at Sears. So far, no foot cancer for me, knock on wood. Nowadays, the person taking the X-ray steps behind a lead shield before shooting the film, and there are generally accepted standards for how much radiation a person can receive on an annual and lifetime basis without incurring undue risk.

X-rays cast a shadow when portions of the beam are interrupted, just as the sun casts a visible shadow of a lobster pot's wooden lattice. That is fine if you are interested in the pot, but what if you are instead interested in imaging the lobster inside? From any given angle, you cannot see the entire lobster. What to do? Walk around the lobster pot and take a photo every 30 degrees—one o'clock, two o'clock, and so forth, knowing that the lattice will obscure part or all of the lobster from some positions. Nonetheless, you can later accurately estimate the lobster's missing contours and dimensions by combining the images. Take the pictures from over 100 evenly spaced positions around the pot and you can begin to understand the concept of computed tomography—CT scanning—also known as computed axial tomography—CAT scanning.

This imaging method came about in the 1970s and rewarded Godfrey Houndsfield, working in England, and Allan Cormack, working in Massachusetts, with Nobel Prizes in medicine in 1979. CT scanning became practical with the advent of high-speed computers, which process the X-ray images taken from all the angles and construct images of the area of interest unobscured by overlying structures. At first it took hours for the computer to acquire the raw data and create images. Now both image acquisition and

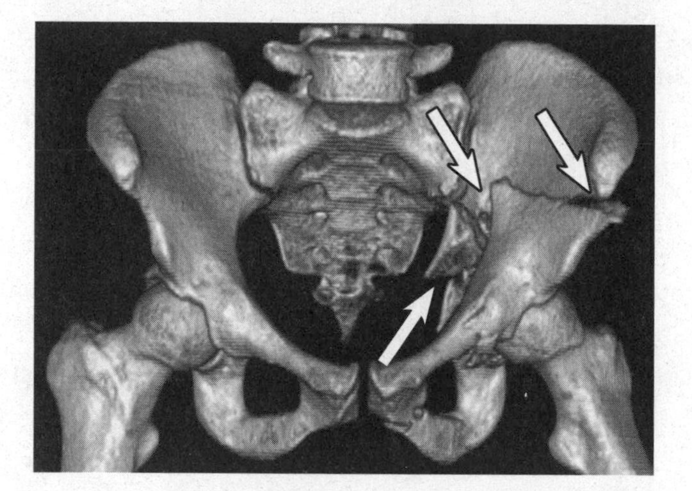

*This three-dimensional image of a fractured pelvis consists of 300 two-dimensional computed tomographic images stacked together. The arrows indicate multiple fracture lines that would be difficult to see and interpret on plain X-rays.* DAVID A. RUBIN, MD, ST. LOUIS, MO.

processing occur in seconds. In orthopedics, CT scanning is most helpful in two instances. The first is when a soft-tissue area of interest is surrounded by bone, such as where nerve roots emerge from the spine. The other is when a shattered fracture involves a joint or is in a complex anatomical area such as the pelvis. Computer-generated three-dimensional renderings can help the surgeon visualize the injury and plan a reconstruction. Such images are impressive and helpful, but CT scanning comes at the expense of considerable radiation exposure.

When viewing their plain X-ray or CT images with their doctors, patients often ask, "How do my bones look? Do I have osteoporosis?" Remember two facts. First, osteoporosis means porous, fragile bone that is prone to fracture because it has diminished calcium content. This occurs naturally with aging and is hastened in women after menopause. These days, anybody who is not living in a cave probably knows about osteoporosis and its perils. Not so widely understood is the second fact: normal X-rays do not reveal the presence or absence of osteoporosis. Factors that preclude a normal X-ray from revealing bone mineral density include the amount of surrounding soft tissue and the duration and intensity of the X-ray beam. That's the bad news.

The good news is that DXA bone density scanning, which is a lot easier to say than dual-energy X-ray absorptiometry, can accurately determine the presence and severity of osteoporosis. Two standardized X-ray beams, one low energy and one high energy, are aimed at the same area of bone, typically either in the lower back or hip. These areas are chosen because they wreak the most havoc when they collapse from inadequate calcium support. The low-intensity beam is mostly absorbed by the soft tissues, so subtracting its effect from that of the high-intensity beam leaves the amount of X-ray that was absorbed by the bone. It is a bit like somebody telling you that they can drive to the grocery store in 10 minutes. That observation has much more meaning if they also tell you how long it takes them to walk there.

Despite our best efforts, we cannot avoid radiation exposure entirely. Some comes naturally from the sun and some from the ground. We get more during a plane flight because the thinner air at high altitude blocks less of the sun's radiation. This fact poses a major, unsolved problem for interplanetary flight because of the absence of any radiation-shielding atmosphere en route and because of the impracticality of covering the space capsule with lead sheets. Stay tuned, or maybe just stay earthbound. Even so, a CT scan of your lower back zaps your fragile DNA with about 70 times the amount of radiation from a single X-ray view of your chest; and radiation from a chest X-ray is about the same amount you are exposed to by just hanging out on Earth for 12 days. By comparison, radiation exposure from a mammogram is about 4 times that of a chest X-ray, and exposure from a DXA scan is about 1 percent that of an X-ray.

Most would agree that the benefits of a mammogram, chest X-ray, or DXA scan far outweigh the risks of the radiation exposure. Even an occasional and judiciously planned CT scan may help maintain or restore your health, but avoid suggestions such as, "I don't know what's wrong, so let's get a CT scan." Remember, it took decades for the damaged DNA in the skin of early dentists to turn into cancers. Similarly, avoid being your own doctor and stating, "I would just feel better, Doctor, if you ordered a CT scan."

What about the X-ray scanners at the airport where travelers step into the glass booth and reach for the sky? Is that bad for the body? No. The X-ray units at the security checkpoints are extremely low intensity and monitor backscatter, radiation that bounces back from the surface of the traveler, whereas medium-intensity X-rays for plain films and CT and DXA scans penetrate the body and cast shadows according to the density of the tissues that they pass through. Since the backscatter units image just the surface, the vertical beams that swoosh around make images of travelers' front and back sides but not their insides, where some people harbor total hip replacements or plates and screws. But TSA is not particularly concerned about metal objects that you cannot access during the flight. Sure, the security scanner radiates the skin, but it's about one-tenth of what you receive by just walking around for a day and 1 percent of what a routine chest X-ray delivers.

Sometimes TSA sends travelers through the X-ray backscatter booth and sometimes through the metal detector that looks like a doorframe. The walk-through metal detector sends a pulsed magnetic field across the archway approximately 100 times a second. At the end of each pulse, the magnetic field reverses polarity and collapses at a speed recognized by the circuitry. If a metal object passes through the archway, it retards the speed of the magnetic field's collapse. The alarm goes off. Maybe.

The size of the metallic object has some bearing. Dental fillings slip by, but sometimes so will a 1-pound hip implant, depending on its composition. The plates, screws, and wires used for fracture fixation are traditionally made of stainless steel, which is about two-thirds iron. Some newer plates and screws are made of titanium. Total joint replacements are often made from stainless steel or from an alloy of several elements. Each of these metals has its own magnetic signature, which means that some are more likely than others to set off the alarm. Additionally, the detectors themselves have variable sensitivity; high, of course, when travelers are late. (Hint: Moseying or racing through the archway reduces the detector's ability to identify metal.) The wands are more sensitive than the archways, partly because the pulsing magnetic fields can be brought closer to your body and to any concealed metallic object.

The bottom line is this. If you are a reincarnation of Evel Knievel (who, according to the *Guinness Book of World Records*, sustained and survived 433 fractures) or have had total knee and hip replacements and a spinal fusion, prepare yourself for the wand and a pat down. With less metal, it depends on the metallurgy and the mood of the machine. If the alarm goes off, the agent will wand you and wish you farewell.

One other main category of bone imaging also involves radiation, but in this instance the radiation comes from a short-acting radioactive element rather than from X-ray beams. Known as bone scanning, and entirely different from DXA bone density scanning, this imaging technique entails injecting a molecule that is attracted to osteoblasts, which are particularly abundant anywhere they are at work rapidly forming or remodeling the skeleton. Hence, the injected substance finds its way to osteoblasts, regardless of why they are working hard. Then a Geiger counter performs a total-body survey and identifies any hot areas where osteoblasts have taken up the chemical because a cancer, infection, or fracture is present, even if the condition has not progressed to the point of being visible on X-rays. The bone scan cannot say what is prompting osteoblast activity, but it focuses the doctor's attention to the diseased area(s) in order to plan further investigation with other imaging studies or a biopsy. Take, for example, a patient with prostate cancer, which in its late stages may spread to bone. Rather than taking total-body X-rays, which would be costly and entail considerable radiation exposure, a bone scan would light up any area of involved bone, whether in the skull or little toe. The amount of radiation exposure from a bone

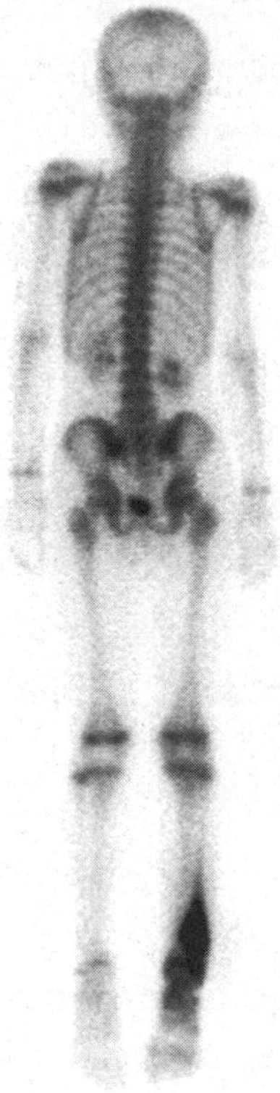

*This bone scan of an adolescent viewed from behind shows increased uptake of a radioisotope in the right lower leg, indicative of an actively growing bone tumor. Symmetrical areas of isotope uptake at the knees, shoulders, and wrists characterize normal bone growth.* David A. Rubin, MD, St. Louis, MO.

scan is about the same as that received from a CT scan of the lower back—far less than that received from a total-body CT scan. Consequently, bone scans provide a comprehensive screening. Total-body bone scans also are valuable in detecting child or elder abuse, where multiple fractures of different ages would be suggestive.

To avoid the risk of radiation altogether when imaging bones, consider the merits of magnetic resonance imaging. The development of MRI followed quickly on the heels of CT scanning and led to a knighthood for Brit Peter Mansfield and Nobel Prizes for him and American Paul Lauterbur in 2003. Initially, MRI was called nuclear magnetic resonance, but some people were twitchy about the word *nuclear*, even though this form of imaging does not use any radiation. The MRI machines are huge and hugely expensive. In essence they are gigantic magnets, strong enough to make the subatomic particles in water molecules wobble. When the magnet turns on and off rapidly, highly sensitive receptors detect faint electronic signals created by the agitated particles. Computers then interpret these minute messages to create an image of the area being wobbled. In one common display mode, tissues that contain lots of water (such as fat) appear white. Tissues with less water in them, including bone, are dark.

MRIs can identify problems in soft tissues that X-rays would pass right through without creating a shadow. MRIs are particularly useful when an organ without calcium—like the brain or heart—is enclosed in a bony cabinet, such as the skull or rib cage, and where the overlying bone would obscure images of the innards. MRIs can also help diagnose ligament, cartilage, and tendon problems in shoulders, hips, and knees, where large amounts of bone can shroud important soft-tissue details from X-ray analysis.

Although MRIs avoid the issue of radiation exposure, they have other safety issues. The magnets are massively strong and have caused metal chairs and oxygen tanks to fly across the examination room with disastrous consequences. The same is true for metal inside the body. It is a bad idea to get an MRI if you have a metallic ear implant, a cardiac pacemaker, a total joint replacement, or a plate and screws. Extensive studies on experimental animals and on humans indicate that MRI imaging has no deleterious effects on the body. Nonetheless, current guidelines recommend use of other means of imaging for women during the first three months of pregnancy.

The worst issues of MRI, however, are not related to physics but to psyche—one during the examination and one after. I will do "during" first because it can be managed with drugs if necessary. Even as powerful as the MRI's magnets are, they have to be placed close to the body in order to make its protons wobble. This necessitates poking patients into narrow tubes and telling them to hold absolutely still while the noisy magnets clang off and on merely inches away. Even the thought of it makes me claustrophobic, which is why some adults and certainly most children require sedation in order to obtain good images.

The other MRI issue could possibly be managed with drugs, but most folk take a dim view of sedating entire populations, so we must fall back on education. As with the high-tech allure of CT scanning, some doctors may order an MRI when other avenues of investigation have led nowhere. Likewise, some patients ask or even insist on having an MRI, thinking that anything short is an incomplete investigation and suggestive of medical incompetence. To the contrary—indiscriminate use of advanced imaging, either CT scanning or MRI, is not likely to find the needle in the haystack but rather find some absolutely normal straws that just do not look quite right. Then a diagnosis of "incidentalitis" leads to anxiety, more imaging, and other forms of testing, biopsies, and treatments for perfectly normal findings. These should have remained unidentified and ignored. Such overdiagnosis leads to overtreatment with myriad risks, both medical and monetary. Education to understand the limitations of MRI is paramount.

Ultrasound is another entirely different technique for imaging the musculoskeletal system. Just as you can holler across a canyon and then hear your echo, high-frequency sound waves bounce off our soft tissues and bones. Different tissues have different rebound characteristics, so the machine produces a reflected image of the anatomy. When properly applied, ultrasound is safe even for a developing fetus, so there are no issues of radiation exposure or claustrophobia distress that are associated with other imaging techniques. Ultrasound cannot penetrate bone; rather, it just bounces off the surface, which limits its application to surface tumors or infections. Orthopedists mainly use ultrasound to image muscles, tendons, and ligaments, but those details are part of another story.

The final way to image bone is to look at it or take pictures of it through a microscope. This, of course, necessitates taking bone out of the body, either as a biopsy or autopsy. Scientists have been looking at all sorts of things under the microscope since it was invented over 350 years ago, but bone has always posed special problems because of its hardness.

Any substance to be examined with light microscopy has to be cut thin enough for light to pass through it. This means that for biological tissues, the slice has to be just one cell layer thick. For soft tissues, this entails embedding the specimen in a block of paraffin or ice, shaving off a gossamer thin layer, and mounting it on a microscope slide. But bone would destroy the shaving blade immediately, so special tactics are in order. One method is to soak the bone in acid for several weeks until the calcium is dissolved and embed what remains in paraffin, then slice away. Another is to embed the hard bone in equally hard epoxy and slice it with the mother of all planing machines, which is so rigid, heavy, and sharp that its blade shaves bone the same as butter. The third way is to grind the bone specimen until it is thin enough to transmit light. For all three methods, specific chemicals added during preparation will stain various components of the specimen differently and allow for identification of subcellular components, including collagen and cell nuclei.

Conventional microscopes use glass lenses to enlarge and focus light images on the examiner's retina or in a camera. The microscope you used in high school or college could magnify objects 500 times, which can provide a great view of osteoblasts and osteoclasts. For a closer look, scientists turn to electron microscopes. Magnets focus a beam of electrons, which either pass through or get blocked by portions of the specimen, and create an image. Since electrons have a much shorter wavelength than does visible light (think high-pitched screech versus rumbling thunder), magnification can be nearly 1,000 times greater than what a standard light microscope can provide. For scanning electron microscopy, electrons bounce off the surface of three-dimensional objects and can provide stunning images, such as seen on p. 60.

Transmission electron microscopy, like light microscopy, provides only two-dimensional images but can home in on much smaller objects, including individual collagen fibers. Is it possible to peer into the future and understand what is in store for living bone?

*Chapter 9*

# THE FUTURE OF CONCEALED BONE

CONSIDERING BONE'S 500-MILLION-YEAR HISTORY and understanding some of the ways that doctors and other scientists have probed its mysteries and nudged it for special purposes, bone certainly has had a grand past and deserves our utmost respect. But what does the future hold?

Two emerging technologies whose impacts on bone specifically and musculoskeletal health in general are at the opposite ends of a size spectrum. On a population scale, artificial intelligence is beginning to sort through billions of bits of health data and imaging pixels and identify patterns where investigators presently see only chaos. The time will come when artificial intelligence will be able to say with assurance that a 47-year-old Caucasian weight lifter who smokes should have one procedure while an identically aged man of differing demographics should have another. At the other end of the spectrum, nanotechnology, working at the level of atoms and molecules, will impact bone growth and repair. For instance, investigators, endeavoring to enhance maintenance and healing, are beginning to add submicroscopic ceramic and metallic particles into bone's half-billion-year success story.

Another area, easier to visualize and ripe with interest and fascination is limb regeneration—tweaking the body's cells and molecular messengers to grow a missing arm or leg. Some salamanders do it, even multiple times. Also, some lizards regenerate their tails after a predator snacks on the original one. For both salamanders and lizards, the amputation stump seems to go through the same developmental sequence that occurs in embryos; but here it is merely a limb, not a whole body. Probing the cellular and molecular mechanisms responsible for such amazing regenerations occupy the energies of numerous investigators. One day they may be able to molecularly tickle the amputation stump of a human's ankle or thumb and get it to grow a replacement. I wouldn't hold my breath for this advance, however. Let's look instead at some advances that are closer to fruition.

There is an increasing need for better bone, particularly as more people live longer. Of course, advancing age predisposes bones to osteoporosis, and countering osteoporosis with exercise may lead to activity-related injuries, including sore backs, broken bones, and irritated tendons. These are situations that sound like decades-long orthopedic job security. Maybe not. Many improvements in bone health will come from advanced understanding of biology rather than from improved surgical techniques and implants. This should not surprise us, because as we learn more about bone on cellular and molecular levels, modulations will be directed at the same level. After all, from a cell's or a collagen fiber's perspective, surgery is gigantic, crude, and primitive. Rather, cells and collagen molecules would likely rather respond to chemical agents far tinier than themselves.

Medications to prevent and treat osteoporosis will become more effective and have fewer side effects. Imagine the day when a 95-year-old tennis player falls and dents the court rather than breaking his hip. We will also see custom antibiotic treatment. The medications will be formulated to destroy the particular strain of invading bacteria while taking into account any metabolic idiosyncrasies of the patient. Consider this analogy. You could stop a rogue computer by unplugging it and disabling its auxiliary power supply. Or knowing its operating system and passwords, you could craft several lines of code and destroy the offending hard drive. The same will be the case for custom-designed antibiotic therapy.

Before DNA was discovered in the 1950s, certainly nobody anticipated gene therapy, yet it is in use today, although not yet for bone conditions. It is a terribly (and maybe wonderfully) complex endeavor that can change one's genetic code. Research has focused on treating inherited lethal diseases, many of which you have never heard of, and you certainly do not want to see any of them listed in your medical record. The researchers identify the faulty code in the patient's DNA, clip it out using special enzymes, and replace it with correct code from a plant or other animal. (If gene therapists ever read what I just wrote, they might balk at my extreme oversimplification, but I am writing about bone, and I leave the intricacies of their story to them.)

When I think about how controversial genetically modified foods are, it is a brave new world indeed to contemplate the time when the molecular biologists can tweak our genes to have stronger bones, tougher cartilage, and hearts that are not such prima donnas.

A far less controversial topic and one with great potential is tissue transplantation, which includes organ transplantation and, especially for orthopedics, limb transplantation—procedures that replace a missing or damaged part using a similar part from another person. I include this topic as a biological advance rather than a technical advance, because even though the task is time-consuming and tedious, any skillful hand surgeon can reconnect the bones, vessels, nerves, and tendons of an amputated hand back onto a forearm of its original owner. Once the circulation stabilizes over the next few days, both the replanted part and the owner thrive.

This transfer of a kidney, heart, liver, or limb from one person to another is far more complicated, but not because of the operation. The surgical techniques required for moving body parts from one location to another are well established. The hat trick necessary for success in cross-body tissue transplantation is biological rather than surgical—persuading the recipient's immune system to befriend the transplant and treat it as self, or at least as friend, not alien.

The problem is that the immune system is constantly on the alert for invaders. It objects vehemently when confronted with viruses, bacteria, rose thorns, bee stings, and other enemies. The system goes into overdrive when it detects foreign materials and does everything possible to exterminate them. Moving bone grafts around from one site to another in the same individual is no problem, because the body recognizes the transferred cells as self, just in a different location. Not so with somebody else's cells.

Starting about 50 years ago, kidneys were the first organs to be transplanted. Immunologists worked overtime to dose recipients with strong antirejection drugs required to trick their immune systems into accepting the transplanted organs. It worked. The recipients' immune systems, however, were not only blasé about the new kidneys; they were equally indifferent to other invaders. Therefore, transplant recipients had lifelong difficulties defending themselves against infections and cancers. Nonetheless, soon the benefits of kidney transplantation outweighed the risks, especially when the immunologists gained experience with the optimal doses of the least toxic anti-immune medications. Kidney transplantation became routine. Heart transplants followed shortly thereafter, then lung, liver, and other vital organs. Yes, the antirejection drugs pose life-threatening risks of their own; but if facing early death without a new vital organ, probably most of us would opt for the transplant and accept the risks.

What about a missing hand? Although useful and aesthetic, its absence is not life-threatening, which leads to a big question. Is it appropriate to transplant a limb for an improvement in *quality of life* and risk the life-threatening complications resulting from the required immune suppression? True, hands, faces, uteruses, and penises are being transplanted, and the sponsoring medical institutions are ballyhooing each triumph loudly. Not so well reported are the failures or complications, including failures stemming from the recipients' altered immunity.

Why am I carrying on so long about a pie-in-the-sky treatment as dangerous and controversial as limb transplantation? Because in due time, the immunologists will find a balanced treatment that will allow recipients to befriend the transplant immunologically *and* maintain adequate immune competence. Presently, there are rare individuals who get a free pass for limb transplantation. One example is transferring a hand onto a person who is already immunosuppressed because that individual previously received a vital-organ transplant. Yep, that patient's immune system is already blasé, so there is no great added risk from attaching some more alien parts. The other instance is when the potential recipient has an (extremely generous) identical twin. Since identical twins' immune systems are the same, transferring parts causes no rejection issues at all. Those readers who are identical twins, take caution. Be nice to your sibling. You never know when you might want them to help out. Likewise, be wary if your gimpy twin seems overly interested in the condition of your knees.

How about just engineering a new organ rather than transplanting one from somebody else? Scientists have been cloning patients' skin for years, particularly for burn victims who need vast amounts of replacement skin. For this purpose, a surgeon harvests the patient's own skin cells from unburned patches. In the laboratory, a technician nurtures the skin in a nutrient broth where the cells grow and divide many times over. Then the cells are attached to a porous, biodegradable film and laid on the patient's burns. Rejection is not an issue because the cells are the patient's own. These cultured cells restore the skin's barrier to infection and water.

In recent years, large cartilage defects in the knee are eligible for similar treatment. The surgeon takes small plugs of cartilage from noncritical areas on the margins of the patient's knee joint and sends them to the lab for reproduction. A month or so later, the surgeon seals a membrane over the offending pothole and injects into it millions of the great-, great- . . . great-grandchildren of the original cartilage cells. The new cartilage cells join together and restore a smooth surface to the joint.

Can identical cell culture techniques work for bone defects? How nice it would be if we were able to replace a shattered, infected, or cancerous bone with a new one that had been fabricated in the lab from the patient's own cells. Tissue engineers are experimenting with such advances, but bone poses several problems not encountered with laboratory-produced skin or cartilage.

The skin needs only a simple and temporary scaffold—the porous biodegradable membrane. The cartilage cells do not need a framework; rather they are merely suspended in liquid and injected into the covered defect. Bone cells, however, need a rigid three-dimensional framework that will resist bending, twisting, and compressing forces. Moreover, the framework has to be permeated with channels for capillary ingrowth. If the channels are too small, the capillaries cannot gain access to the interior of the framework to nourish the osteoblasts. If the channels are too large, the scaffold will likely be weak and fail.

Bone cells pose another problem. Our bodies need millions of osteoblasts, and a patient cannot spare enough donor bone to provide the needed amount to populate the scaffold. Rather than trying to grow them in the lab, biologists look to stem cells, which are like utility baseball players—second base one day, catcher the next, depending on the team's changing needs. A fertilized egg is the ultimate stem cell. It divides repeatedly, and its offspring over time differentiate into heart muscle, nerve, skin, bone, and all the other amazing cell types that constitute a newborn baby. Stem cells slowly disappear as we age, so embryonic stem cells would be the best ones to coax into becoming osteoblasts if it were not for the ethical issues surrounding such use of fertilized human eggs.

There are several intriguing alternatives. Although only small numbers of stem cells can be harvested from an adult's bone marrow or circulating blood, more can be found in fat, which abounds in modern Americans. Accordingly, a quick liposuction not only trims the figure but can also provide stem cells to be engineered into whatever tissue is in short supply. An alternate source that does not require skin penetration is baby teeth. Wait until one falls out, fend off the Tooth Fairy, get the tooth on ice, and hustle it to a lab that, for a price, stores it in their deep freezer. Then, perhaps decades later and on the unlikely chance that the tooth's owner requires some tissue engineering, tooth-derived stem cells are available. How fortunate, just pop the vial out of the freezer and hop to it. Of course, this means that the cell bank is still in business and that its freezer has never gone on the fritz. Stem cell banking sounds zany to me, but if I ever need some of these utility players, I will regret having traded my baby teeth to the Tooth Fairy for a quarter each.

If we have some stem cells that the scientists can coax into converting to osteoblasts and we have a scaffold, are we done? Hardly. At least three hurdles remain for successfully engineering bone. The first is persuading the cells to stick to the scaffold and to migrate into its innermost reaches. The second is inducing capillary ingrowth to provide the osteoblasts with food and beverage. Finally, the osteoblasts need to receive chemical love letters from pituitary, thyroid, and testes or ovaries in order to prosper, divide, and make new bone.

Tissue engineering is getting a boost from another emerging technology—additive manufacturing. Also known as 3-D printing, it is not only revolutionizing industrial fabrication, it is also contributing to engineered tissues. Layer by layer, investigators are printing all the necessary cells and frameworks for artificial kidneys, livers, and hearts. Similar production of engineered bone lags behind because of its rigid nature. But when the techniques are perfected, just imagine if you need a replacement bone. Simply take your baby tooth out of the freezer, extract its stem cells, take them to your local photocopy shop, and fire up the 3-D printer, or maybe even a 4-D printer. Here a three-dimensional object is printed and then alters it shape, origami like, either over time or with the application of heat, moisture, or light. Such an implant could be inserted through a minuscule incision and, once inside, change shape to accomplish the reconstruction.

Even though engineering living bone is still on the horizon, additive manufacturing already has real applications in orthopedic surgery. Using data obtained from an MRI or CT scan, a 3-D printer can produce a full-scale plastic replica of a shattered bone. Sitting at her desk, the orthopedist can then study the sizes and shapes of the fracture fragments and plan appropriate fixation. Such 3-D modeling is particularly useful for weirdly contoured areas, including the elbow, pelvis, and heel. Not only can the surgeon leisurely study the fracture from all angles, she can also pre-bend plates and determine the lengths of the necessary screws to maximize surgical efficiency. Such full-scale models will also be valuable learning aids for students and patients.

In the near future, additive manufacturing will also produce custom hardware for fixation of difficult fractures. And when unique situations preclude the use of off-the-shelf components, 3-D printing will also fabricate custom bone or joint replacements. In the past, such one-off parts could be machined by conventional techniques, but it took weeks. Patients with broken wrists or ankles can also anticipate getting a cast or brace printed in the orthopedist's office while they wait. A surface scanner will take pictures of the injured ankle from multiple angles and send them to a 3-D printer, which will print a custom, formfitting mold in the patient's preferred color. The brace will not only hug the puffy ankle effectively and comfortably but will also avoid putting pressure on any sore spots.

Both off-the-shelf and custom orthopedic plates and screws will come in an array of nonmetallic materials, including carbon fiber. It is lightweight, strong, and radiolucent, meaning that X-rays pass right through without the hardware casting a shadow. Such implants will allow for full visualization of the fracture, even though covered with a plate.

Even wilder than custom, radiolucent implants are plates and screws that do their work and then disappear. Consider for a moment the history of plate-and-screw composition. Just over 100 years ago, surgeons were using whatever screws and pins they found in their workshops or in their spouses' sewing boxes. Unfortunately, these

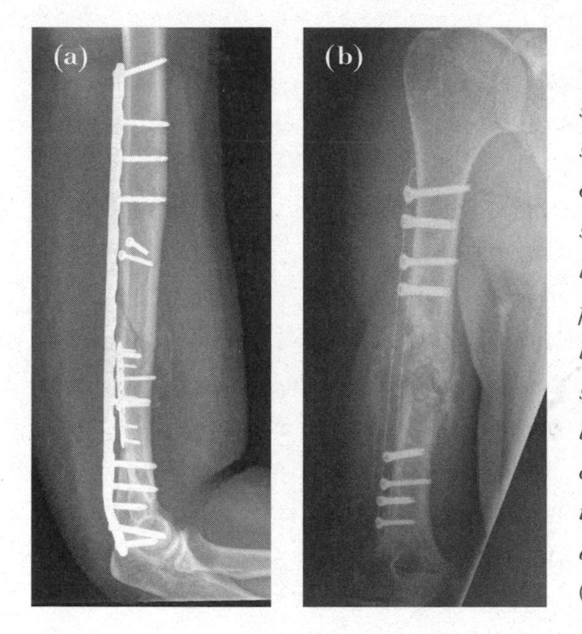


*(a) Conventional stainless steel plates, one long and one short, span multiple fractures of this upper arm bone. (b) A similar fracture is bridged by an innovative carbon fiber plate, which is secured to the bone with conventional metal screws. The plate does not block the X-ray beam and is only evident by the fine steel wire, which indicates its borders.* (A) CLIFTON MEALS, MD; (B) ALIDAD GHIASSI, MD.

borrowed bits of iron or aluminum corroded quickly in the body's saltwater environment. Along came stainless steel and orthopedists were off and running—plates and screws galore. Titanium hardware became available more recently and has some advantages, previously discussed. But how about plates and screws that could secure a fracture long enough to ensure bone healing and then dissolve?

Initially after fracture fixation, the implant would need to resist all bending and twisting forces. As the bone began to heal and the plate began to slowly disappear, the bone and plate would begin sharing the mechanical loads. Piezoelectric forces generated by the (controlled) stresses would stimulate the cutting cones to strengthen the bone. The plate would gradually render itself unnecessary before disappearing entirely. For decades, investigators have been stirring up various resins, some derived from cornstarch, and forming them into plates and screws. The research has yet to find a formula that combines sufficient strength, minimal bulk, low tissue reactivity, and adequate duration of rigidity. When such a material arrives, surgeons can 3-D print custom plates while they wash their hands and don their gowns and gloves in preparation for fracture fixation surgery. Compared with metal plates and screws, secondary surgery for hardware removal would be unnecessary. Rather, the body would tote the bioabsorbable fixation away one molecule at a time.

Presently, every hospital where orthopedic surgery is performed has shelves and cabinets filled with boxes and trays of orthopedic hardware. Both in stainless steel and titanium, plates range in length from less than 1 inch to over 12 inches. Screws in the same materials have various thread configurations and range in length from an eighth of an inch to over 4 inches. Some of this array may never be used, but it is necessary to have the entire set available just in case. One 3-D printer using a suitable bioabsorbable material would render the storehouse of metal hardware obsolete.

Minimally invasive surgery is another area of rapidly advancing technology. In orthopedics, it started with knee arthroscopy. Make a couple of small cuts in the skin, insert a light saber and TV camera along with some skinny instruments, and voilà, the athlete is back in the game. It is the same for rotator cuff repairs in the shoulder.

Smaller incisions often equate with less pain, swelling, and bleeding and thereby portend a faster convalescence. Consequently, the trend toward minimally invasive surgery will continue. At the same time, however, surgeons' hands are not becoming smaller. We will always have trouble manipulating instruments and tissues in tight spaces, such as inside the pelvis. Enter the robots. Their arms and attached tools are awesomely flexible and compact. They can therefore work efficiently through small incisions and in tight quarters. They do not tremble or get tired. Yet to some nonsurgical specialists' chagrin, robots will not replace surgeons.

Because the robot does not act independently, the accurate terminology is computer-assisted surgery rather than robotic surgery. Sitting comfortably at a computer console in the operating room rather than stooping over the operating table, the surgeon guides the robot with a joystick-equipped controller. The time may come, however, when orthopedists work their magic via augmented reality while sitting at the kitchen table (working from home) or while reclining under a palm tree (working vacation).

Despite anticipated advances in treating bone disorders, both from innovations on the horizon and those not even presently imagined, there will always be job security for orthopedic surgeons as long as people continue to fall off skis, motorcycles, and electric scooters.

Also, skeletal injuries will always require imaging, but one day our current methods of taking pictures of bone may be obsolete. I hope so. X-rays and CT scans expose the body to radiation, which is known to be harmful in large doses. An MRI requires holding still for too long in a noisy cocoon. Ultrasound cannot see inside bone, which limits its usefulness. Pulse-echo ultrasound, however, is among the newer imaging techniques that might one day replace DXA scanning to diagnose osteoporosis with increased accuracy and without radiation exposure.

The ideal bone-imaging technique would not use radiation or otherwise damage the body or psyche and, after the patient holding still for just a second, would provide full-color, three-dimensional likenesses of the patient's innards. Farfetched? Maybe, but who would have anticipated X-rays or MRI in the first place?

Novelists for centuries have anticipated space exploration, and science fiction may soon become science. In the coming decades, lunar colonization and human spaceflight to Mars will become realities. Is it technically possible? Yes. Is the human psyche ready? It seems so. Are human bones up to the challenge? Not yet. Weightlessness poses a big problem. In space, the skeleton does not have to resist gravity, so it tends to melt away. The bone's seemingly unneeded calcium enters the bloodstream. The surplus is more than the heart could ever want, so the kidneys filter it out. But high concentrations of urinary calcium promote the formation of kidney and bladder stones.

Astronauts attempt to counter these adverse effects of calcium mobilization by exercising several hours every day. Weight lifting does not work because the barbells are weightless. Also, running or walking on a conventional treadmill does not work because the astronauts are weightless and launch themselves off the machine with the first step. Instead they use resistance bands to simulate gravity. These are large, stretchy rubber strips that connect a harness on the astronaut's shoulders and hips to the exercise machine. Nonetheless, the astronauts lose 1 to 2 percent of bone mass per month, which is similar to what an elderly man on Earth loses in a year. This means that a six-month stay on the International Space Station causes about a 10 percent loss of bone. Such a rate of bone loss would be unsustainable on a three-to-four-year round trip to Mars.

Shoes containing strong magnets have been considered, but they create more problems than they solve, including their interference with electronic devices on board. Planners have also considered creating artificial gravity by spinning part or all of the space capsule, making it a large centrifuge. Theoretically this would work, but the cabin would either have to be 100 feet in diameter or spin at a rapid, energy-consuming rate, neither of which is presently practical. Furthermore, lunar gravity is only one-sixth of Earth's, and Martian gravity, although stronger, is only a little over one-third of Earth's. Nobody knows whether such forces are strong enough to maintain bone health. Then travelers to Mars may want to take their families along, and the effect of zero gravity on growing bones is entirely unknown. Nonetheless, kids would undoubtedly have a blast doing cartwheels and flips during the whole trip.

The answer to maintaining bone health in space will probably not come through exercise or artificial gravity but through improved medications similar to those used presently to prevent and treat postmenopausal osteoporosis. At some point, harnessing the complex metabolic changes that bears use to avoid bone loss during hibernation could help.

Should extraterrestrial species be discovered, it will be fascinating to discover how they support themselves against gravity. I only claim that bone is the *world's* best building material, not necessarily the *universe's*.

◆◆◆◆

Part One, "Bone Concealed," comes to a close. In life, the world's best building material is seldom seen. We like it that way. In bone's second life, however, the story is different. Exposed bone performs myriad out-of-body roles. Once revealed, bone becomes an excellent recorder of Earth's history and human culture.

*Part Two*

# BONE REVEALED

*Chapter 10*

# BONE LEFT ALONE

SEEMINGLY EVERY NATURAL HISTORY AND anthropology museum in the world displays a skeletal replica of Lucy, our 3.2-million-year-old human predecessor. Her skeleton is far from complete, yet the discovery of her bones in 1974 in Ethiopia and the subsequent study of their shapes have been critical to our understanding of human evolution. Lucy has given us one of the biggest scientific discoveries of our time: first our ancient ancestors stood up and walked, and only later did they develop large brains. Lucy is amazingly intact—an extraordinary specimen that is fully sufficient to determine sex, brain size, and walking posture, and is therefore a priceless morsel of world heritage. To a casual observer, however, this assemblage of Lucy's bones seems anything but. Five pairs of ribs, not 12? Only one finger and one toe? Half a pelvis? Lucy, how could you walk with less than one leg? Overall, paleoanthropologists uncovered only about 40 percent of the skeleton. What's on display is approximately *Lu*. What happened to *cy*?

Taphonomy (from the Greek *taphos*, meaning "burial," and *nomos*, meaning "law") is the branch of paleontology (the study of ancient life) that attempts to understand what happens to a newly dead plant or animal before it decomposes entirely or before its fossil is discovered. When it comes to our ancient ancestors' bones, taphonomists ask, "How did *these* end up *here*, and where are the rest?" Unfortunately, given the capricious effects wrought by the forces of

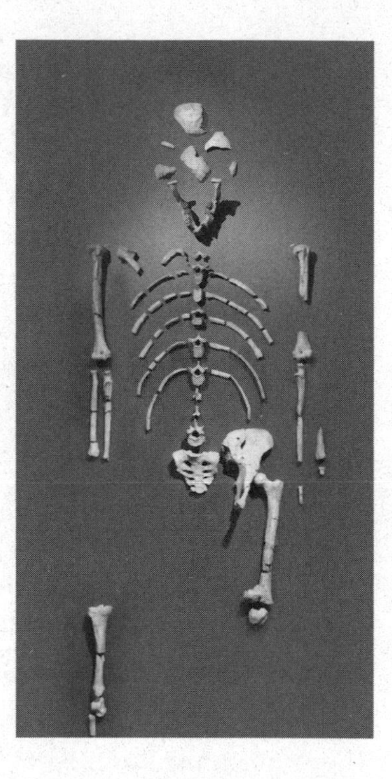

*The skeleton of Lucy, our 3.2-million-year-old human predecessor, is about 40 percent complete. Many museums display cast replicas of this world treasure. The original bones are under lock and key in Ethiopia.* NATURAL HISTORY MUSEUM OF LOS ANGELES COUNTY.

time and nature, the stories that bones tell can be confusing. Part of a taphonomist's job is to solve these riddles.

Bones exposed to air and sunlight decompose just as skin and internal organs do, only much more slowly. Moisture evaporates and the fat inside the bone breaks down within a year or two; surface crackling ensues, followed by flaking and fissuring that deepen over time. Eventually, the bone splinters into brittle bits. If the bone is undisturbed, this decaying process takes between 6 and 15 years, depending on temperature, humidity, light intensity, and the size of the animal. Exposed bones shielded from direct sunlight can last hundreds if not thousands of years, as evidenced by skeletal remains found in caves and rock shelters.

However, exposed bones are rarely left alone, which may explain how bones scatter or accumulate in strange places. Hyenas swallow whole bones and may travel considerable distances before disposing of them. Ravens bring sheep bones into their nests, and years later skeletal parts may rain down in seemingly odd places. Otters carry fish onto stone outcroppings hundreds of feet above the water, where the bones remain for long periods. Primitive humans accumulated bones beside their firepits, and Native Americans stampeded mammoths over cliffs to kill them en masse before driving them extinct. Then they turned to bison. At the foot of one bluff in Alberta, Canada, fittingly named Head-Smashed-In, the jumbled bison bones measure 39 feet deep. At the foot of a similar cliff near Lyons, France, lay the remains of an estimated 100,000 wild horses.

Perhaps the most adept bone accumulators are owls, a surprising source of information for zoologists and taphonomists alike. Unlike most carnivorous birds, owls do not use their beaks and talons to tear their prey into bite-size pieces. Rather, an owl swallows its prey whole, breaking down muscle and fat into a digestible liquid. But bones and fur resist this decomposition. Hours later, the owl regurgitates its leftovers in the form of a compressed, gray-brown, and odorless pellet, about the size of the last two segments of your thumb. Inside these offal treasures are bones important to ecologists studying owl diets, molecular biologists studying DNA, and osteologists studying skeletal structure and function.

Inanimate forces also subject exposed bone to movement and degradation. Ice and flowing water move bones downhill, and wind rolls bones in any which way. Wave and tidal action abrade and polish them and can simulate wear caused by human use. Gnaw marks might indicate a small animal attempting to satisfy its calcium needs. Hack marks suggest humans satisfying their protein needs.

Buried bones don't always fare better and can also mislead taphonomists. Warm soil hastens bone decay, and soil compaction flattens skulls and rib cages. It can also cause long-bone fractures if the legs were crossed at burial and the lower one later becomes a fulcrum. Moisture content and soil acidity can differ considerably over short distances, which may expose part of a skeleton to conditions not experienced by the rest. The highly acidic volcanic soil of Japan and Hawaii accounts for the dearth of bones in their archaeological records.

*(a) An intact barn owl pellet contains the indigestible fur and bones of its prey. (b) A CT scan of the pellet demonstrates a jumble of undigested bones. (c) The dissected pellet reveals most of the bones of one small and three larger rodents.* (B) DAVID A. RUBIN, MD, ST. LOUIS, MO.

Buried bones may become reexposed or shift beneath the soil because of tectonic plate upheaval, mud and ice slides, and even human exhumation and reburial. During an archaeological dig, the small bones of the fingers, toes, and wrists may be overlooked, if they haven't already decomposed completely, which occurs because their outer walls are thin and easily dissolved.

It gets worse for taphonomists and paleontologists working to reconstruct history through bone analysis. Trampling can break bones, creating the illusion of a violent end—though taphonomists can usually tell by the configuration of a fracture whether it occurred during life, at the time of death, or years after. Trampling can also churn and mix bones from one animal together with those from others. Stepping directly on a buried bone can push it deeper into older layers of sediment; conversely, stepping next to a bone can shift it closer to the surface and lead scientists to the erroneous conclusion that it was more recently deposited. When a skeleton is found relatively intact and in a lifelike position, it is likely that silt covered the body shortly after death and sheltered it from movement.

How do paleontologists (fossil fans) and anthropologists (human fans) find these bones in the first place? There are three key ways.

The first is to go where the money is—excavating at known or likely burial sites. These include the Olduvai Gorge in Tanzania, the La Brea Tar Pits in Los Angeles, and cemeteries and Indian mounds everywhere. The second is to search areas that are subject to erosion and exposure, including riverbanks and the ancient seabeds that once covered the central part of North America. Paleontologists will walk for miles along exposed faces of sedimentary rock, their eyes glued to the ground looking for bone; winter weather erosion may have exposed what had been completely hidden last season. The third is serendipity. Building contractors will often unearth ancient remains and then find their construction projects halted while impassioned scientists descend into building foundation sites, utility trenches, and highway cuts to salvage priceless markers of the ancient past.

Although her skeleton is incomplete, *Lu* is one such priceless marker, and a combination of planning and luck led to her discovery. In the 1970s, paleoanthropologist Donald Johanson, of the Cleveland Museum of Natural History, surmised that the Awash Valley region of Ethiopia was a likely repository for bone remnants that would reveal much about our human origins. So he mounted an expedition. During his second season of exploration, in 1974, a gully that he had searched twice before revealed a fragment of *Lu*'s arm bone. It caught the paleontologist's eye, and a thorough search ensued. We are lucky that Lucy's body was silted over shortly after her demise and that it remained undisturbed for the next 3.2 million years. Luckier still, she surfaced at a time when a trained eye was there to spot her.

So where are the rest of Lucy's bones? It is possible that the *cy* portions of her skeleton were previously exposed and eroded or were washed away. Or maybe early on *cy* fell prey to scavenging or other taphonomic processes. Still, the fact that we have *Lu* to document an important step in human development is astounding. Five years prior to her discovery, Lucy may have been completely concealed. A year or two after, a paleontologist searching the same area might have found only *L*. Or nothing at all.

For ease of communication, we often say *bones* when we discuss Lucy or dinosaurs, but what we really have are fossils—stone replicas of the original bones. Fossil formation is a complex interplay of bone size and porosity along with local mineral composition and concentration, acidity, and temperature of the mud covering the skeleton. When conditions are just so, water permeating through a bone's porous surface leaches out a calcium hydroxyapatite molecule, which is then immediately replaced by another mineral molecule dissolved in the water. The bone turns to stone, one molecule at a time, over tens of thousands—sometimes hundreds of thousands—of years. The mud surrounding the bone also turns to stone, but its constitution is different because it did not start out as hydroxyapatite. If the fossilized bone weathers away a little more slowly than the sedimentary rock within which it is embedded, then later—even millions of years later—when a sea, lake, or riverbank dries up, a passerby might spot a partially exposed fossil.

Paleontologists know which rock strata are fossil rich and will search areas where those particular layers are exposed. Even so, they may search for months without finding a fossil. When they do strike it rich, the painstaking and tedious work begins, carefully chipping and brushing away the surrounding stone to extract the fossil.

I get brain freeze when I consider two probabilities. First, only a tiny fraction of animals from millions of years ago became fossilized. Second, most of those that did turn to stone are still buried or have already become exposed and weathered away. The fossils that we do have, therefore, constitute only the tippy tip of an iceberg and are a fortuitous and miraculous insight into Earth's history.

Although fossilization is the most enduring, there are other means of skeletal preservation, which I will describe going from cold to hot. Ice can preserve whole bodies for tens of thousands of years. From time to time, the bodies of woolly mammoths emerge from thawing Siberian permafrost. By far the most famous and carefully studied case of ice preservation, however, is a human specimen named Ötzi, nicknamed the Iceman. In 1991, hikers in the Austrian Alps discovered Ötzi's body in a melting mass of ice; he had met his demise over 5,000 years earlier. Not only his skeleton but also his skin, internal organs, stomach contents, and clothing were preserved. Scientists leaped at the opportunity to scrutinize Ötzi's remains, but once specimens like Ötzi are identified and retrieved, they must be kept frozen to prevent rapid deterioration of the soft tissues. This makes storage, study, and display particularly challenging. Investigations of Ötzi are ongoing and will likely continue in perpetuity as investigators apply advanced methods of nondestructive analysis.

Liquid water rather than ice has trapped and preserved other bodies. Bog bodies are one example. Peat bogs—wetlands that accumulate deposits of dead plant material—typically contain cold water, low oxygen levels, and varying degrees of acidity. These conditions preclude bacterial growth and decay, leading to a macabre form of pickling. Compared with frozen bodies, bodily remains found in peat bogs usually have well-preserved skin and soft tissues *or* well-preserved bone, but not both. When acidity levels are high, the peat bog "tans" the skin and innards and preserves clothing, but dissolves bone; when acidity levels are low, the opposite occurs. The oldest bog body to be discovered, the Koelbjerg Man dates to about 8000 BC and is on display at the Møntergården Museum, in Odense, Denmark. Scandinavia has been the region where most bog bodies turn up. Investigators believe that the Koelbjerg Man drowned, but other bog bodies were likely intentionally disposed of, and many show signs of torture and violent deaths suggested by nooses, stab wounds, and beheadings. Human sacrifice? Criminal execution? Unruly neighbor? Nobody knows for sure.

Provided that bones are in the shade, air and then water can also preserve bone, as evidenced by several recent finds in underwater caves on Mexico's Yucatán Peninsula. The remains of three humans along with other mammals, believed to have fallen to their deaths through unsuspected sinkholes 12,000 to 13,000 years ago, have been found in the region, where porous rock formations abound. Many centuries later, sea levels rose and flooded the caves that entombed the animals, preserving their remains. Cave-diving paleontologists, perhaps taking a break from summers spent in blistering hot dinosaur beds, discovered them thousands of years later.

Ancient human remains occasionally turn up on the seafloor, typically in conjunction with shipwrecks. These include notable discoveries in Stockholm harbor and in the Caribbean and Mediterranean Seas. Other bones found on the open seafloor consist of aquatic animals that died naturally as well as land animals that either drowned at sea or died inland and were swept into the salt water. The carcasses of these animals can drift on the ocean surface for weeks until the soft tissues deteriorate, allowing the skeleton to sink. (A taphonomist described it to me as "bloat and float.") These bones may scatter widely, depending on storms, wave action, and currents, or they may settle close together. For whales, marine biologists call such an accumulation a whale fall.

Amber, which is fossilized plant resin, has proved to be an incredible room-temperature preservative for many life-forms. It is unlikely that an amber boulder large enough to contain a set of human bones will ever turn up, but for studying ancient small creatures, amber is a biologist's treasure trove. Amber nuggets, some dating as far back as the dinosaur era, may contain insects, seeds, and pollen; a few contain tiny frogs, reptiles, birds, and small mammals. Microcomputed tomography can image their skeletons without disturbing the encasing amber, and the detail is astounding. Investigators recently discovered a 2-inch-long baby snake that was entrapped in amber about 99 million years ago. Its 97 vertebrae, each about the size of a sesame seed, can be viewed clearly from any angle. What is equally remarkable is that such tiny, fragile creatures, which have been encased this way for millions of years, have not sustained any degradation during scientific scrutiny—so these specimens should, in theory, be available for continued study over millions of years to come.

Another sticky preservative, in far greater abundance than amber and also warmer in temperature, is asphalt—the thick, gooey component of crude oil. Shallow underground petroleum deposits may allow asphalt to seep to the surface and form pools. A number of such sites, commonly called tar pits, exist in Venezuela, Trinidad, Azerbaijan, and several locations in California. Dust and leaves may camouflage the asphalt pool, or a thin layer of water may cover the asphalt, whereupon unsuspecting animals wade in, become trapped, struggle, and die. In the process, they may attract predators, which then meet the same fate. The good news for paleontologists is that these pits are densely filled with plant and animal material ripe for study. The bad news is that the bones are jumbled due to terminal struggling and trampling. Nevertheless, tar pits provide scientists with a wealth of information, not only about vertebrate animals from long ago, but also about insects and seeds of the same vintage. At the La Brea Tar Pits in Los Angeles, a short bike ride from where I live, most of the retrieved animal skeletons are between 11,000 and 50,000 years old. This is far too recent to reveal any dinosaurs, but the displays at the adjacent Page Museum include extinct dire wolves, saber-toothed tigers, mastodons, one human, and other fascinating creatures. Outside, faintly odiferous asphalt continues to ooze to the surface, but fences surrounding the pools prevent cyclists from riding across. Imagine a paleontologist a thousand years from now finding my bicycle.

The last preservative to be discussed is hot. In AD 79 Vesuvius erupted and quickly smothered everything and everyone in nearby Pompeii—first with superheated, hurricane-force wind chased by volcanic ash that became 80 feet deep. Following the disaster, survivors abandoned the entire region, and only in 1748 did a surveying engineer rediscover the city. Archaeologists began digging and continue today. The unique combination of deadly heat and sudden burial in a dry, oxygen-poor environment allowed for excellent preservation of wooden structures, works of art, and yes, bones. Some bodies were buried in ash that later solidified, and the encased soft tissues eventually turned to dust. Complete skeletons remained in the ash's body-shaped voids. There is no telling how many of these voids diggers unwittingly destroyed before archaeologists in the nineteenth century recognized these hollows for what they were and began preserving them by filling the cavities with plaster upon discovery. After the plaster solidified, they removed the surrounding ash. This left a cast of each body and even impressions of its clothing. In recent years, computer tomographic imaging studies have revealed the well-preserved bones inside the intact "plaster bodies."

Herculaneum, near Pompeii, was spared the first of Vesuvius's blasts, allowing time for most of its residents to flee. Later that same day, however, fumes as hot as those that had swept over Pompeii caught several hundred people, who died instantly from the heat. Archaeologists found the largest assemblage of bodies in a boathouse, where local residents, unable to escape, apparently huddled together in a futile attempt to escape the intense heat. In the following hours, vast rivers of volcanic mud swept through Herculaneum, engulfing the city with deep layers of what was essentially quick-drying cement. This preserved charred bones but without the surrounding cavities found in Pompeii. Although universally lethal, the heat in neither city was sufficiently intense to cause bones to crumble. Scientists have found the remains, unintentionally preserved and left alone until discovery, a valuable albeit macabre resource. What can be learned from intentionally interred bones?

*Chapter 11*

# DEFERENCE TO BONE

An argument began in 1908 regarding which species was the first to bury its dead and for what purpose. Anthropologists had just unearthed a nearly complete 100,000-year-old Neanderthal skeleton from a cave in France. They determined that a grave had been dug and that the body, carefully placed in the fetal position, was then buried. Because excavation and documentation techniques at the time of the skeleton's discovery were crude, skeptics howled from the outset. Today, the truth remains elusive. And even if the burial was intentional, was it to protect the body from predation, or was it done for spiritual purposes? The latter possibility raises issues about the origins of abstract thinking, which some anthropologists believe the Neanderthals did not possess and argue that mortuary practices have always been uniquely *human*.

Specialists further challenged the advent of abstract thinking in 2013 with the discovery of a new human-like species from deep inside a South African cave. Access was difficult for the investigators, and it must have been equally so for all time. Reaching the burial site entailed climbing, crawling, and squeezing through an 8-inch gap, all in the dark. The principal investigator could not fit, so he employed small, lithe, female anthropologists with caving experience. They found numerous fossilized bones from at least 15 individuals who spanned a wide age range and who had died roughly

300,000 years ago. None of the bones showed any signs of predation or violent trauma. Investigators have concluded, though not unanimously, that this was a burial site. The skulls of these human cousins are small. Were their little brains therefore capable of the symbolic thought, language, and ritual behavior necessary to trundle their deceased kin, with apparent difficulty, into the cave? Fortunately, the bones have survived to raise these questions. Unfortunately, the bones alone do not solve the mystery.

Reported in 1981 and beyond dispute, the oldest purposely buried *human* remains come from a cave in Israel. Numerous pieces of ocher—a red, iron oxide pigment—rested near the remains. Five skeletons, now all approximately 100,000 years old, were orderly arranged, two with deer antlers in their hands—so-called grave goods. Based on this discovery, we know that for at least 100 millennia humans have been paying deference to their deceased kin, whose bones have survived to document funerary practices and, by extension, symbolic thinking, a respect for the dead, and possibly the consideration of an afterlife. The means and methods of respecting human remains—some perhaps merely for the practicality of eliciting closure and clearing the air of noxious odors—vary widely and reflect human cultures.

Several ancient groups have tucked their deceased into rocky alcoves on imposing, overhanging cliffs, sometimes hundreds of feet aboveground. These people left no written history before moving away or dying out, so the motivation for arduously transporting dead bodies to these precarious locations is unknown. For the Tellem people in what is now Mali in West Africa, these alcoves, containing thousands of skeletons, a scattering of grave goods, and some remnants of rope, are all that remain of their existence from roughly 900 years ago. The extreme inaccessibility of the site leaves it undisturbed and virtually unstudied. From approximately the same era, the Chachapoya lived high in the Andes of northern Peru and also housed their dead in cliff crannies, although they mummified them first. The Chachapoya encountered the Spaniards in the sixteenth century and succumbed entirely to European diseases. Unfortunately, opportunists have plundered most of these gravesites, so the remains are of little use to anthropologists.

Coffins suspended hundreds of feet high are visible on cliffsides scattered across southern China, Indonesia, and the Philippines. Dating back at least 3,000 years, wooden boxes or hollow tree trunks, balanced on cantilevered stakes, project from the cliff faces or protrude from crevices and cavities hewn from the rock. The Toraja people of Indonesia continue this practice today. The original rationale for these hanging coffins is unclear; it may have been to prevent plunder and predation, to spare the use of valuable farmland below, or to allow the deceased to be closer to the spirits above. Were these heavy coffins, some containing sand along with the body, lowered from above or raised on scaffolding from below? The oddity of and fascination with these vertical cemeteries have made them tourist attractions.

Less precariously situated but still aboveground, tree and scaffold burials were common among Native Americans; drawings and descriptions of the practice indicate wide use ranging from Quebec, Nebraska, and Wyoming to the Pacific Northwest and Alaska.

Secured with leather bindings, robes and blankets contained the newly deceased along with his most treasured possessions. The packet then rested in a tree or on a high platform built from branches. Only men received tree burials; the bodies of women and children were left in the brush, subject to predation. After a year or two, the "elevated" remains, some of which had fallen to the ground and all of which were no longer of interest to bears and wolves, were buried.

Whereas most cultures take pains to avoid predation, some welcome it in their burial practices. In so-called "sky burials," Buddhists in Tibet and in equally rocky and treeless neighboring countries carry the bodies of the recently deceased high into the mountains and leave them for the vultures. This custom is practical where it is hard to dig graves and where wood to fuel cremation is scarce. It is also spiritual, as Buddhists believe that a person's spirit moves on at death and that the corpse is an empty container, which other living creatures can and should share. Once the vultures have picked a skeleton clean, the bones are hammered and mixed with grain and milk and fed to smaller birds. Some of the bones, however, may be recycled into ritual musical instruments and vessels, another way that the living pay deference to bones.

Trumpets made from human thighbones and drums crafted from human skulls are integral to traditional Tibetan meditation rituals—ones that stress the fleeting nature of life and material existence. The kangling (literally, leg + flute) is a human thighbone with the hip end cut off. A circular opening in the bone's shaft becomes the mouthpiece. The large flares at the knee are left intact, except for two gouged openings through which air and sound escape once the bone's spongy interior has been removed. The shaft of the bone is thick, hard, and durable; but the bone thins out near the knee and is brittle. This explains why the "bell" end of a kangling is covered with tightly sewn skin, sometimes human, or with thin metal sheeting. The covering adds needed durability because generations will use this venerated instrument that would otherwise crumble away.

The reverent energy imbued in the kangling is believed to derive from the spirit of the bone's original owner—so bones from individuals who were free from worldly faults are favored. These include children, youths with clear minds, and monks and nuns with unbroken vows. Likewise, a kangling derived from a saint or sage would have great "power of realization" and would be able to channel human energy.

Human skull drums, known as damarus, are part of the same rituals and share symbolism and energy with the bone trumpets. Damarus are constructed from remains of a man and woman carefully chosen for their life attributes. Their skullcaps, joined at their poles and attached to a stick, have mantras inscribed in gold on their interiors before skin is stretched across the openings. The players of these drums hold the stick and rotate the drum horizontally by twisting the wrist. Large beads strung on cords swing around and

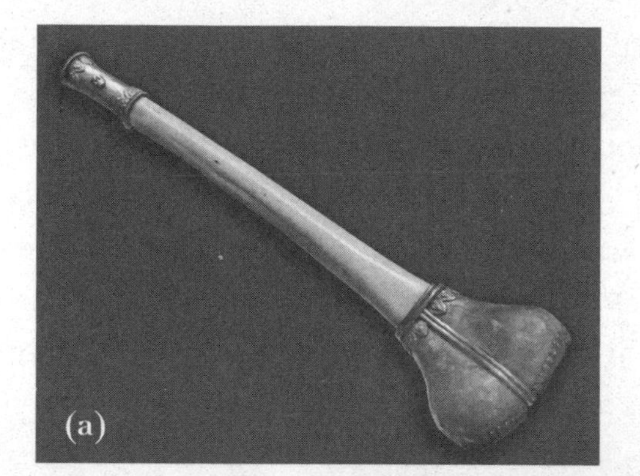

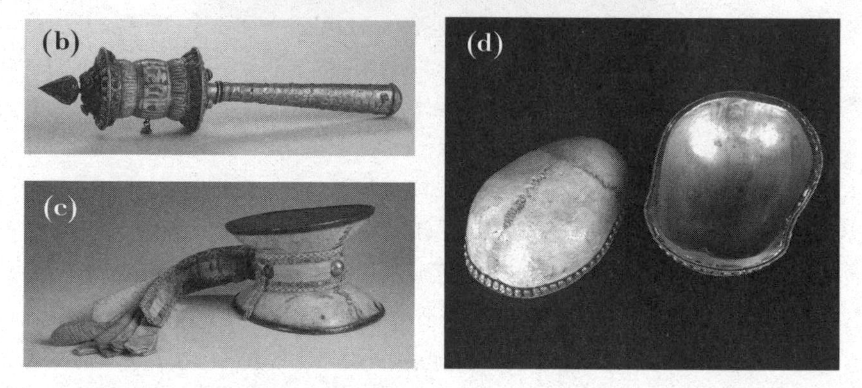

*These bone objects are integral to Tibetan meditation rituals: (a) thighbone trumpet, or kangling; (b) prayer wheel with bone central cylinder, silver caps, and handle inlaid with gemstones; (c) drum, or damaru, made from two skulls; (d) skullcap bowl, or kapala. In (c) and (d), the tight joints where the skull bones interdigitate are evident.* (A) BRITISH MUSEUM; (B, C) LOS ANGELES COUNTY MUSEUM OF ART; (D) BONE ROOM.

strike the drum heads. A practitioner may play a damaru in one hand and either a kangling or bells in the other.

Individual human skullcaps may also become bowls, or kapalas, and are also central in Buddhist and Hindu rituals. Ornately carved or decorated with precious metals and gems, the kapalas have various meanings and uses. These include serving as a vessel to offer sustenance to wrathful divinities and as a meditation aid to remind the faithful of life's impermanence.

Intentional disposal by burning dates back 42,000 years. The body of Mungo Lady, so named for the Australian lake where a scientist found her in 1968, had been burned; then her bones were crushed and burned again. In a charred state, they were covered in ocher, the same pigment used 60,000 years earlier and one-third of the way around the world in Israel, the site of the first known human burial. Since Ms. Mungo's time, various cultures and religious groups at different times have favored cremation. It has also served a point of aspersion between them. Hinduism prescribes cremation, Islam forbids it. As a space-conserving measure, Japan presently mandates it. Judaism has traditionally disapproved of the practice, but some reform groups support it. Cremation was common in ancient Greece and Rome, and the practice waxed and waned in Europe until Christianity swept across the continent. Early Christians, believing in resurrection of the body, were adamantly against it (unless, of course, it was used for burning an adversary at the stake). Within several hundred years, the turn away from cremation was so pronounced that anthropologists have traced the advance of Christianity across Europe through the emergence of cemeteries.

Traditionally, cremations were open-air funeral pyres, and there was little expectation of salvaging any of the remains beyond a few scorched bone fragments. Native American tribes in some parts of Florida, however, spared the skull from the fire and used it to contain the cremains.

In the late nineteenth century, cremation ovens began appearing in Europe and the United States, and the practice slowly gained popularity. Today, cremation ovens are heated to 1,600 to 1,800 degrees Fahrenheit—hot enough to make iron glow cherry-red and to entirely vaporize the body, except for the skeleton. The bones become extremely brittle and break apart but may still be identifiable as such. A powerful high-speed blender, wonderfully known as a cremulator, can then pulverize the remains. The resulting substance, commonly referred to as ashes, is in fact the amazingly durable mineral component of bone. Weighing on average about

5 pounds, a human's gritty remains are worthy of familial respect and, if not scattered, usually reside no more than a fireplace mantle's height above Earth's surface.

Perhaps the most commonly practiced method of respecting bodily remains is also the oldest, and it takes us underground. In addition to deference to those who have passed on, burial reduces the spread of disease and foul smells and, in some cultures, prepares the departed for an afterlife. In favorable soil conditions, buried bones can fossilize and therefore remain available for discovery and analysis many thousands of years later. From Shakespeare's pen and King Richard II's lips come "What can we bequeath save our deposed bodies to the ground?"

When bracelets, beads, pottery, and other grave goods accompany skeletal remnants, scholars can link the bones to the socioeconomic status of their owners and to the beliefs and rituals of their cultures. The knowledge gained from exhumed skeletons has been profound—as are ethical issues that until recently many have widely ignored. While scholars and opportunists alike have clamored for some combination of knowledge, fame, priority of discovery, and wealth, unearthing buried bodies has often come at the expense of families and peoples who believed they were putting their kin to rest with deference and love forever.

Not all buried bodies were lovingly laid to rest, however. Archaeologists have discovered at least three 7,000-year-old graves in central Europe, each containing tens of skeletons that show signs of blunt trauma and that are unaccompanied by any grave goods. Humankind's inhumanity apparently began early. Single body gravesites can also be perplexing. Discovery of so-called aberrant or deviant burials has shed light on violent, vindictive, fearful, and superstitious tendencies to which humans have at times succumbed. Scholars can only postulate the motives when they find a skeleton buried with an extra set of legs or bodies buried facedown. Some discoveries reveal a skeleton shackled in leg irons or handcuffs or held down with metal stakes or heavy stones. Punishment was a possible motive for decapitating a body and placing the head between the knees or for burying it vertically, head down. Perhaps it was instead a measure to prevent these bodies—possibly thought to be vampires or witches—from returning. Most of the discovered aberrant burials have been located in Europe.

Ranging across all continents except Antarctica, bodies abound that have been respectfully preserved by dehydration in the sand of hot, arid climates. After ancient peoples inadvertently discovered this method of natural preservation, some started intentionally replicating the process, beginning about 7,000 years ago in Chile. The Egyptians started almost 2,000 years later. They considered mummification to be an imperative step for a recently deceased and privileged person to gain admittance to the afterlife. The Egyptians also mummified cats in great numbers along with a scattering of dogs, baboons, birds, crocodiles, and at least one gazelle. Embalmers removed the internal organs, packed the body with spices, and surrounded it with a mixture of salts. After 70 days, they wrapped the dried body in multiple layers of linen strips and placed it in a wooden case, sometimes to be discovered thousands of years later. The bones inside these mummified bodies remained concealed until noninvasive means were applied—first X-ray imaging and more recently computed tomography. Tutankhamun, perhaps the most famous mummy of all time, received such a study in 1986. Images revealed a thighbone fracture that, as complications set in, may have contributed to his demise several days after injury.

Tutankhamun included, royal Egyptian mummies rested in elaborate, permanently sealed underground tombs. Each pharaoh arranged for his own eternal digs. But in other cultures, underground crypts have been more like cities of self-storage units. Such catacombs created for the purpose of mass interment exist in multiple European cities, including Paris, Budapest, and Lisbon, as well as in Peru and the Philippines.

The largest network of catacombs surrounds Rome. Ancient Roman law prohibited burials within the city, but few early Christians living in Rome were able to afford plots outside the walls to bury their dead, and they did not participate in the widespread pagan custom of cremation. Rather, from the second century on, networks of tunnels were dug into the soft volcanic rock outside the Roman city walls. A body would be placed in a side chamber, which was sealed with a stone slab inscribed with the name, age, and death date of the occupant. Many of those Christians so interred had been persecuted and martyred, a practice that continued until Rome officially accepted Christianity in the fourth century. Although no longer persecuted, many Christians still wished for eternal rest in close proximity to the martyrs, so catacomb interment continued for at least another hundred years, enough time for an estimated 500,000 to 750,000 bodies to be so treated. The catacombs slowly fell from use and were eventually forgotten entirely, to be rediscovered by chance in 1578. These catacombs, rich with frescoes, have been studied and visited extensively ever since and are an important part of early Christian history.

But how do the Roman catacombs relate to bones? The connection lies in veneration of the saints, the Reformation, and the Counter-Reformation. In Catholic doctrine, venerating the remains of a saint ensures that the saint will intervene between God and mortals requesting favors of him. The "relic," such as a bone or piece of clothing belonging to the saint, thereby became a touchstone for gaining blessings. These relics were stored in, well, reliquaries, which were containers constructed from precious metals and decorated with gemstones. Relatively small in size, the reliquaries were carried in processions on the saint's feast day, and some had a clear glass window through which the saint's bone was visible. Others were shaped to indicate which bone was contained within. The devout made pilgrimages to venerate the relics, spending money along the way and donating to their saint's church on arrival.

Every church wanted a relic, but the saints had only a finite number of bones that could be divvied up. By the Middle Ages, there were many more bones under veneration than the saints could

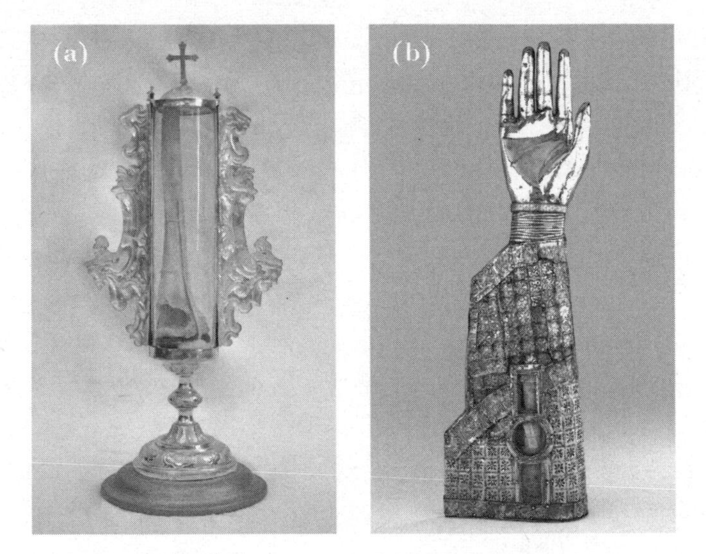


*(a) In the cathedral at Palma, Mallorca, a reliquary displays a leg bone of Saint Pantaleon. (b) The silver hand and gilded copper sleeve indicate that this reliquary for Saint Lawrence contains one or more of his hand bones.* (A) CATEDRAL MALLORCA; (B) BPK BILDAGENTUR / KUNSTGEWERBERMUSEUM, STAATLICHE MUSEEN, BERLIN, GERMANY / ART RESOURCE, NY.

have possibly provided. Reformists Martin Luther and John Calvin objected to the ruse. These relics, many of them fake, had started out as devotional aids but had become false idols. As Protestantism gained ground in northern Europe, the Reformists destroyed reliquaries and recycled the precious materials. Then in 1563, the Catholic council struck back by upholding the tenet that relics were an integral part of religious practice. The council went on to reaffirm the ability of saints to intermediate between God and the faithful, but it set strict guidelines regarding authentication of relics and stated that those without confirmation should not be offered for veneration.

With many of the ancient relics destroyed and the provenance of others in doubt, demand for the real deal vastly outweighed the supply—that is, until 15 years later, when the Roman catacombs were rediscovered. Herein rested the remains of bona fide martyrs along with hundreds of thousands of other early Christians, who conceivably could have been persecuted and thus qualify as martyrs. The devout went wild. Churches and upper-class families sought the martyrs' skeletons as statements of devotion and prestige. Only some of the slabs sealing the crypts were intact and could accurately identify the occupant as a martyr, but all the skeletons sent out from Rome had accompanying documents attesting to their legitimacy. It's unclear whether their martyrdom was based on facts or mere assumptions. Unnamed martyrs received new identities—often the names of popular saints or of local benefactors—to enhance their appeal to the faithful. Other unidentified skeletons received virtuous names, including the Latin equivalents of Constancy, Clemency, and Happiness. In desperation for legitimacy, others were named St. Incognito and St. Anonymous.

The reliquaries from pre-Reformation times contained just a single precious bone or bone fragment. Now, the catacomb saints' whole skeletons offered possibilities for large and elaborate displays, which the churches went to great lengths to realize. Since the "martyrs" had lived in the early Christian era, the new stewards of these complete skeletons dressed them as Roman-era soldiers of the time, bedecked the assemblages with jewels, both imitation and real, and then displayed the spectacle prominently. Miracles were attributed to the catacomb saints. Pilgrims came in droves. The faithful gained solace, with the reminder that such glory was reserved for the truly faithful. The sponsoring churches and monasteries made fortunes.

This business in piety continued for over 200 years and then slowly declined. In 1782, the Hapsburg emperor demanded destruction of relics that could not be fully authenticated. Around the same time, progressive Catholics, influenced by the Enlightenment, wanted their church to be more modern and cultivated. Theft was also a constant concern. Catacomb saints, previously displayed proudly, gradually disappeared from view. Was dispersal of the bones from the Roman catacombs principally a contrived business or a salve for the faithful? You decide.

*The dramatically adorned skeleton of Saint Pancratius, a catacomb saint, is on display at the Church of Saint Nikolaus, in Wil, Switzerland.* KATHOLISCHE PFARR- UND KIRCHGEMEINDE WIL, KIRCHE ST. NIKOLAUS, SWITZERLAND.

Whereas the Roman catacombs were burial sites from which remains were removed centuries later, the opposite happened in Paris. There, bones from individuals buried in the city's churchyards slowly accumulated and eventually created storage problems. The solution was to move the old bones into a labyrinth of under-city tunnels. At the time, it was expedient. Later, the Paris catacombs became a business. (I will describe my firsthand customer experience in Chapter 13.)

Other old cities had similar bone storage problems but managed them differently. Rather than just stacking old bones in catacombs, some churches began displaying them artfully but without any allusions to saintly connections. At St. Michael's Chapel in Hallstatt, Austria, the "bone house" displays 1,200 skulls on shelves. Sorted by family, painted labels on half of the skulls contain the owners' names along with birth and death dates. Other churches began using bones for interior decorative elements. These "bone churches" exist in England, Spain, Poland, the Czech Republic, and in at least three cities in Italy. My wife (reluctantly) and I (enthusiastically) visited one such bone chapel in the lovely medieval town of Evora, Portugal, about 90 miles east of Lisbon. On approach, the ornate stone and stucco facade of the main church gave us no clue to what lay inside. Our first hint was the inscription in marble over the chapel's side entry, which translated, "Our bones that are here await yours."

In sixteenth-century Evora, the Franciscan monks confronted two problems. Storage space for old bones was sparse. Their parishioners were also living the high life because gold was pouring in from Brazil, then a Portuguese colony. To make a statement regarding the transient glories of worldly existence, the monks collected at least 5,000 skeletons from local churchyards and crypts and used the bones to decorate the walls and ceiling of a chapel. In some areas, they pressed arm and leg bones into plaster so that the length of each is visible. In other areas, the bones are viewed end on. Skulls liberally punctuate the geometric wall patterns and define the edges of the columns and arches. Throughout, there is a symmetrical and repeated contrast of dark and light. Strange as it may seem, I found

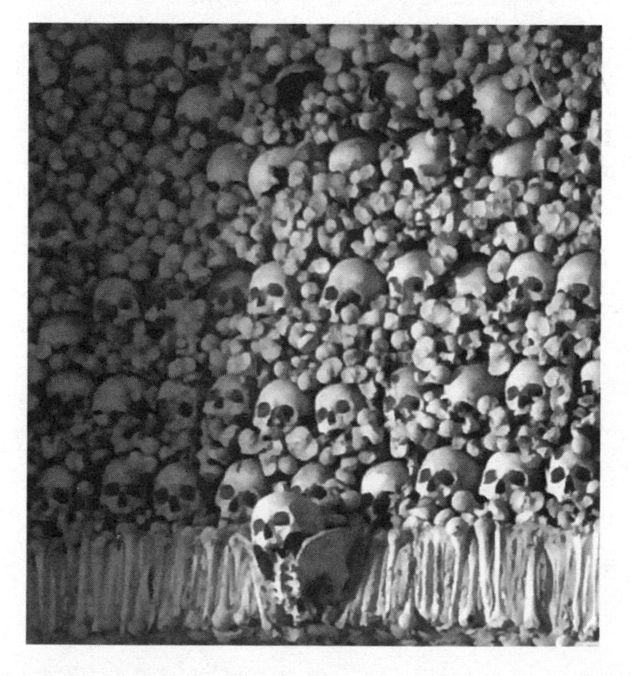

*Dating from the sixteenth century, the walls of this bone chapel in Evora, Portugal, are lined with the skeletal remains of over 5,000 individuals to remind us of life's transience.* Chapel of Bones, Evora, Portugal.

*Indigenous Australians (Aluridja people) are "pointing the bone" at an evil adversary in a long-established assassination ritual.*


the chapel starkly beautiful. My wife politely said she would wait outside while I finished taking pictures.

Through the ages, humans have also deferred to bones in ways other than paying respect to their kin. Take, for example, a practice of indigenous Australians that dates back at least 40,000 years, to the time of Mungo Lady, the earliest known human to be cremated. Other evidence suggests that her kin had been in Australia at least twice that long. With such a lengthy heritage, it should not be surprising that the people developed strong beliefs. These included the notion that death, except for the elderly, was unnatural and that an enemy's evil spell could hasten its arrival. Once a tribe identified the spell caster, which might take years, a designated assassin set forth with a ritualized, dagger-length, needle-shaped piece of bone, be it human, kangaroo, or emu. After stealthily approaching the accused, the assailant chanted a short curse while pointing the bone at the spell caster and then returned home to burn the weapon. The assailed, struck with fear, would become listless over the following days to weeks and then die, as though this "spear of thought" had caused a physical wound.

Perhaps less dramatic but similarly influential was the way an ancient Chinese dynasty used flat bones—ox shoulder blades and turtle shell bottoms—to predict the future. Beginning almost 3,500 years ago and lasting for about two centuries, the royal house of Shang used these bones to glean answers to critical questions regarding crops, military expeditions, hunting, weather, travel, sickness, and the king's health. Small pits were drilled into one side of the bone, and pressing questions were carved into the opposite side with a sharp knife. A hot poker placed in the pit would cause

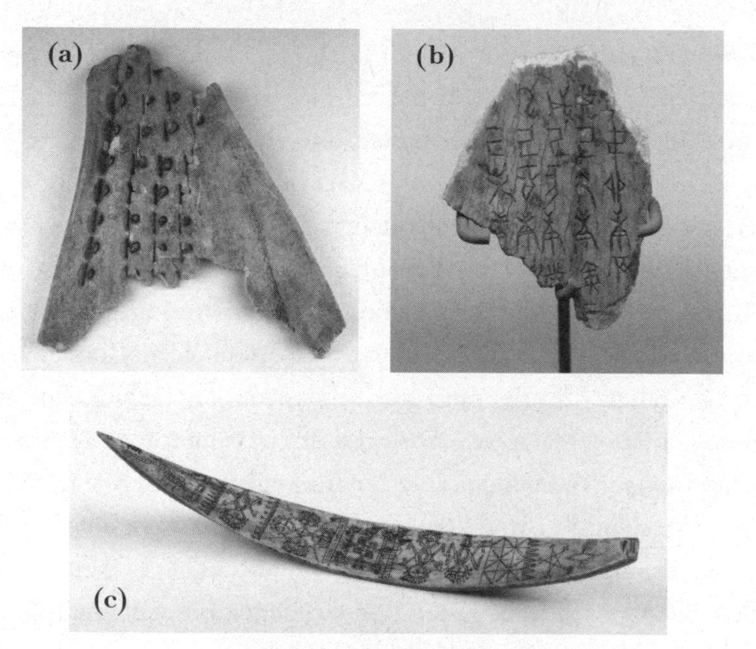


*Divination devices. (a) A hot poker applied to an ox shoulder blade created pits and cracks for the diviner to interpret. This object is from the Chinese Shang dynasty, sixteenth to tenth centuries BC. (b) Also from the Shang dynasty, the writing describes the divination's answers. (c) An inscribed water buffalo's rib from Sumatra provided calendrical advice and medical formulas.* (A) BRITISH MUSEUM; (B) SHANG MUSÉE MARIEMONT; (C) WELLCOME COLLECTION.

the heated bone to crack, and the orientation of the cracks yielded answers, which a diviner or at times the king himself interpreted. Each revealed answer was then inscribed on the bone for future reference. Thousands of years later, these ancient relics frequently turned up in plowed fields. Farmers rubbed away the inscriptions on what they believed were the bones of dragons, ground up these former notepads, and sold them to be ingested for medicinal purposes. In 1899, an astute antiquarian recognized ancient script on such a bone fragment and pursued its provenance. The markings, sharply inscribed on the bones, are the oldest existing example of Chinese writing and reveal much about the culture of the time.

Once the significance of these "oracle bones" became clear, a brisk collector's trade ensued, including the foisting of fakes. Excavated from Chinese soil that had proved favorable for long-term storage, 200,000 legitimate oracle bones, mostly fragmented, have been saved from a medicinal fate. A quarter of them bear inscriptions that use about 6,000 different characters, 2,000 of which have modern linguistic equivalents, so their meanings are known. (Most of the others are proper names.) In addition to answering questions, the bones recorded details of celebrations, calendar events, tax increases, and genealogy. Inscriptions regarding astronomy include the earliest documentation of a solar eclipse and of a passing comet. Overall, the inscriptions give a remarkably detailed view of the Shang people's lives. We should be thankful that dragon bones were not panaceas.

In Sumatra, priests inscribed the wide rib bones of water buffalo. They were passed from generation to generation and contained calendar information used to divine the best days for rituals or travel along with formulas for magic and medicine. The beauty and poetry of the inscriptions were also valued, as was the good luck that the bones conferred.

Inhabitants of southern Africa used smaller rectangular sections of flat bone with various symbols and shapes carved into them as divining tablets. Likely even older is the custom of tossing unmodified "knucklebones" and interpreting the meaning of their landing

*(a) A South African "bone thrower" foretells the future using knucklebones, likely from a goat or sheep. (b) Carved and inscribed bones also served as divination devices in South Africa.*
(A) Wellcome Collection; (B) Science Museum, London.

positions. Not only fortune-tellers but also gamers used these almost cube-shaped anklebones that came from goats and sheep. They were the precursors to dice.

Good luck associated with bones was a theme throughout ancient Europe. Almost 3,000 years ago the Etruscans, who lived in what is now central Italy, believed that birds, especially geese, could foretell the future. Among other rituals, the Etruscans would sun-dry the V-shaped bone from a goose's neck and preserve it in hopes of retaining some of the fowl's magic powers. They would stroke the bone and make a wish—hence the "wishbone." The Romans adopted many Etruscan customs, but in the course of seeking good fortune, they squabbled over wishbones and broke them, from whence comes the expression "lucky break."

Romans took the wishbone custom with them to England, where by 1607 the locals called a wishbone a "merrythought" because fortune would favor the contestant owning the longer end with an early marriage. The Scots quantitated this prediction by drilling a hole through the flat part of the wishbone and then placing it over the bridge of a girl's nose. The number of tries she took to successfully pass a thread through the hole was how many years it would be until her marriage. (Perhaps extreme nearsightedness was considered a favorable trait in a wife.) When Europeans settled America, they transferred the fortune-telling capabilities of old-world fowl to new-world turkeys.

Voodoo came to the New World via West African slaves. One enduring ritual is casting chicken bones for telling fortunes. Each bone has an individual meaning, as does the location where it lands with respect to the others and to the circle into which it is cast. But even before it was called a New World, shamans of indigenous Caribbean tribes performed a ritual using bone spatulas and cohoba, a psychoactive powder. This custom entailed pulverizing seeds of a local tree, mixing the powder with tobacco, and inhaling the mixture to induce a hallucinogenic trance. The ritual was preceded by fasting and purging with the aid of a vomiting stick made of bone. Then the shaman prepared himself to receive the "pure food" of the spirits, who would bring him messages and allow travel to the supernatural world. Cohoba made everything appear upside down, reversed from side to side, brightly colored, and ever changing. The handles on the vomit spatulas reflected this otherworld with fantastic upside-down figures and fierce, imaginary animals.

Also for spiritual reasons, Native Americans, particularly in the Great Lakes region, placed dream catchers—hoops supporting a fiber web and strung with sacred items, including bone—over their infants' cradles as protective charms. The Inuit tribes of the Pacific Northwest inscribed sections of bear thighbone and plugged the ends with cedar bark to contain lost souls. The living believed that these "soul catchers," worn around the neck, held spiritual powers, as did many other bones that they held dear. A walrus penis bone, for example, supposedly imparted the powers of the animal onto the

*The vomiting stick, made here from a manatee bone, facilitated the fasting and purging required to enter the trance of the cohoba ritual, practiced by shamans of indigenous Caribbean tribes.* Museum of Native American History.

human owner. Native Americans also made flutes out of an eagle's wing bones and could simulate the bird's call. In so doing, the revered eagle became a ceremonial figure, and at times other eagles would appear in response to the flute's sound. It may seem strange to us, but hunting cultures venerated the animals they killed, thanked them for their lives, and memorialized them with drawings, carvings, and musical instruments, some of which are extant, and rituals, most of which are lost.

In great contrast to the ways that humans have deferred to hunted animals and bone over time, Lucy died unceremoniously, perhaps by falling out of a tree unnoticed; yet her fossilized bones are now treasures to be forever treated with deference. Human remains that were more recently discovered indicate in early humans a capacity for abstract thinking and spirituality; we can tell this because the remains were managed with more respect initially. But deference to deceased fellow humans has been by no means universal. Thus, bones, typically the only enduring human tissue, answer some questions about life and death while raising many others.

*Then Sampson said, "With a donkey's jawbone I have killed a thousand men" (Judges 15:16).* SCULPTURE BY GIAMBOLOGNA, ABOUT 1562. VICTORIA AND ALBERT MUSEUM.

Fortunately, though, not every lesson derived from bone is ponderous and morbid, or even deferential. What is deference for one culture might merely be reference for another. Therefore, without intending any disrespect, I offer here a sampling of more worldly references to bone across time.

Three from the Old Testament of the Bible stand out. From Genesis: "And the rib that the Lord God had taken from the man he made into a woman." In Judges, Sampson recounts, "With a donkey's jawbone I have killed a thousand men." Ezekiel wrote that God had set him in the midst of a valley that was full of bones, and "there was a noise, and suddenly a rattling; and the bones came together, bone to bone." All three of these allusions to bone have been immortalized multiple times in sculptures and paintings, but only the latter in song, which is often the first anatomy lesson children receive. "Toe bone connected to the foot bone. Foot bone connected to the heel bone. . . ."

Dia de los Muertos is a three-day Mexican holiday that honors saints and remembers the faithful departed. The celebratory and

*The Day of the Dead celebrations began in Mexico and have spread around the world. In Los Angeles, depicted here, festivities include parades and stage shows with costumed dancers.*

commercial aspects now often eclipse the original intent of these autumnal holy days, and the party atmosphere has spread around the world fueled by an opportunity to poke a finger in the Grim Reaper's eye. Bones abound, for they are the most enduring remembrance of lost souls. Skeletons dance and parade. Face-painted celebrants indulge in bone-shaped cookies and skull-shaped candies.

Over the centuries and around the globe, nothing symbolizes death more than a human skull, easily recognized and evocative of fascination and fear along with remembrance. In the visual arts, tombstone carvers, sculptors, painters, and tattoo artists have all

*Skull imagery has lost its impact as a symbol of danger or death as evidenced by the entrance to the Vortex Bar and Grill in Atlanta and by its depiction on a wide array of consumer goods.*

joined in. Skulls, with crossbones added just in case the message needed emphasizing, have appeared on military insignia, pirate flags, and bottles of poison. Today, skull imagery is everywhere and has mostly lost any emotional impact. These allusions include Alexander McQueen's high-end designer scarves and now even house slippers, coffee containers, and beer cartons. (I can show you my skull socks.)

Not to be outdone by visual allusions, Camille Saint-Saëns composed, in a minor key of course, *Danse Macabre* for orchestra, in which a xylophone's plinking well represents bones rattling. He used the same instrument to produce bone imagery in the *Fossils* movement of his *Carnival of the Animals*.

In the theater, the most famous skull is Shakespeare's Yorick, the court jester of Hamlet's youth. Over 400 years later, Tchaikovsky tried tagging onto Yorick's immortality by bequeathing his skull to the Royal Shakespeare Company. For several months, the great composer's skull played Yorick; but alas, once its identity was known, those in charge feared that the skull was upstaging the actors, so it was shelved.

Some names also refer to bone, although the connection is not always obvious. For instance, Golgotha (from the Arameic) and Calvary (from the Latin) mean "the Place of the Skull" and refer to the site near Jerusalem where Jesus was crucified. Similarly, Calaveras County, California, popularized in Mark Twain's tall tale, received its name from skulls (*calaveras* in Spanish) found along a local riverbank. Another river and its adjacent city, this time in Australia, are named Brisbane in honor of an early governor. The family name is rooted in Old English, where *brise* meant "to break" and *ban* meant "bone," but I have been unable to discern whether this family suffered broken bones, broke the bones of others, or set broken bones. In other instances, bone's lesson is clear.

*Chapter 12*

# BONES THAT TEACH

THE MATERIAL EXAMINED IN PALEONTOLOGY, the study of ancient life, is remarkably old, a minimum of 11,500 years. The scientific discipline, however, is relatively new compared with its academic parents, geology and biology. Several reasons account for its start, albeit slow, as a field of study in the 1600s. At that time, Renaissance collectors of stone and stone-like objects called them fossils (from the Latin for "things that have been dug up") without considering that some of their finds had once been alive. Two philosophies contributed to that misconception. Followers of Plato maintained that there could be associations between living and nonliving objects such that they could resemble one another. Aristotle's followers supported the idea that seeds of living organisms could enter the earth and grow to resemble plants and animals. Leonardo da Vinci, Mr. Renaissance himself, disagreed with both philosophies because he surmised the biological origin of fossilized sea creatures that he discovered high in the Italian mountains. Others basically ignored or scorned his observations.

In the following century, Robert Hooke, whose expertise spanned astronomy, mathematics, physics, and biology, suggested that some fossils represented organisms no longer in existence. Extinction, however, was a concept that most of his contemporaries were unable to accept for either philosophical or religious reasons. The concept of extinction gained credibility a century later when observers noted distinct skeletal differences between the remains of elephants and long-frozen mastodons whose previous existence was indisputable.

Beginning in 1808, the discovery of large reptile fossils provided clear evidence that giant reptiles had once stalked the planet. This stirred excitement in the scientific community and led to the coining of the word *paleontology* to describe the discipline. Throughout the remainder of the nineteenth century, discovery, observation, and analysis of fossilized bones proceeded with fits and starts.

One of the fits related to an eccentric early disciple. This was William Buckland, who in 1822 published the first complete description of a dinosaur (although the word, meaning "terrible lizard," would not be coined for another two decades). Many scientists discounted Buckland's findings based on his quirky personality. Among other idiosyncrasies, Buckland performed his geology fieldwork in a cap and gown, often lectured from horseback, and brought his pet bear to meetings; and after extensive experimentation, Buckland attempted to classify animals on taste alone and claimed to have eaten his way through the animal kingdom. Mole and bluebottle fly he found distasteful. Fellow paleontologists were loath to dine at Buckland's home.

Until the mid-nineteenth century, scientists of all disciplines were wealthy, curious gentlemen who supported their own explorations and investigations. They often displayed an array of minerals, fossils, stuffed exotic animals, antiquities, and art in *cabinets of curiosity*, which were actually rooms—*wonder rooms*. Such a collection established and maintained its owner's rank in society, and the rooms became salons where like-minded intellectuals and nobility gathered.

With respect to fossils, this amateurs-only approach changed as museums, universities, and government surveys began employing geologists and paleontologists. Their discoveries proved popular for both cultural and economic reasons. The findings, including the comparative anatomy of long extinct animals, provided evidence to support the validity of evolution. Furthermore, awareness of where fossils were found aided the exploitation of Earth's minerals, especially the coal that fueled the Industrial Age.

Dinosaur bones, however, mainly interested only scientists until British artist Benjamin Waterhouse Hawkins turned up in Philadelphia in 1868. He was already a prolific and well-known natural history illustrator for Darwin and other mid-century biologists. He had also created a sensation by fabricating life-size clay reproductions of dinosaurs for a London display. This led to a second large commission, which was to develop an extensive Paleozoic museum in New York City's Central Park. The intention was to dynamically display dinosaur fossils that had recently been discovered in America.

Because New York possessed neither the necessary fossils nor the paleontological expertise for the project, Hawkins ventured to Philadelphia. There he received support from the Philadelphia Academy of Natural Sciences to assemble the skeleton of a 30-foot dinosaur, most of the bones of which were in the academy's collection. He made plaster reproductions of missing bones and creatively supported them all with a metal framework such that the reconstructed skeleton assumed a lifelike, standing posture. This had been done with human bones in ancient Alexandria, but those bones were light and easy to wire together. Hawkins was the first to scale this up to display fossilized dinosaur bones weighing many tons.

The public clamored to see this assembled dinosaur skeleton. Attendance at the academy doubled over the next year. The associated wear and tear on the museum forced the academy for the first time to charge admission. Nonetheless, the display piqued the public's interest in dinosaurs, an interest that remains strong over 150 years later.

Today, no natural history museum seems complete without an assembled skeleton of a large dinosaur posed in a threatening stance, usually in the museum's main atrium, where display height is not an issue and where shock and awe are maximal. By far the most famous of these is Sue, the *Tyrannosaurus rex* who for decades was ready to gobble up arriving visitors at Chicago's Field Museum. Hers is quite

*Benjamin Waterhouse Hawkins is pictured with the phenomenon he created in 1868—the first assemblage of a dinosaur skeleton.* HTTPS://COMMONS.WIKIMEDIA.ORG/WIKI/FILE:HADROSAURUS_FOULKII.JPG.

a story, including a custody battle and the longest criminal trial in South Dakota history, all of which I will describe later.

Fossils, both from dinosaurs and from other vertebrates, both on public display and in stored collections for scholarly research, are mostly the result of paleontologists, both amateur and professional, spending untold hours walking with their eyes fixed to the ground. Fossils slowly "weather out" of ancient rock formations, and these exposed surfaces are usually in regions that are presently arid and hot. Scrambling up crumbly rocky cliffs in withering heat and squatting for hours with a dental pick in one hand and a paintbrush in the other are not for everyone. Most paleontologists, however, seem to thrive on such adversity. A few have also thrived on controversy, as attested by the antagonists in the Bone Wars.

These battles were sordid clashes between two paleontology giants—both of them egomaniacal, ambitious, jealous, and rich. Author Url Lanham delicately describes the adversaries, Othneil Marsh and Edward Cope, this way:

> *At a level above the ordinary garden variety of malicious gossip is genuine hate, which probably is one of the most valuable forces*

*in existence for producing, quick, accurate, incisive, and original thinking. Both Cope and Marsh enjoyed the benefits of this emotion to an unusually high degree.*

Their first meeting, in 1863 in Berlin, was friendly. Marsh was continuing his studies there after having graduated from Yale, where he would later return. Cope had dropped out of school at 16, but by the time of their meeting he had already published 37 scientific papers, compared with Marsh's 2. Cope was quick-tempered and impulsive. Marsh was quiet and methodical. Both were testy. Marsh's landlady said that getting to know him was like "running into a pitchfork." Cope continued throughout his career to do fieldwork in the vast fossil beds between the Mississippi River and the Rocky Mountains. Conversely, Marsh spent four seasons in the field and otherwise paid for others to bring fossil specimens to him at Yale—an armchair paleontologist.

Cope came from a wealthy family in Philadelphia. He lived and supported his research on a substantial family inheritance. Marsh came from a poor family in Lockport, New York. He benefited, however, from the largesse of his wealthy uncle, George Peabody, who was a merchant turned banker turned philanthropist extraordinaire. The Peabody name still rests on multiple educational and scientific institutions in the eastern and southern United States. Among those he funded were two at Yale—the Peabody Museum of Natural History and the Peabody Chair in Paleontology, conveniently occupied by his nephew. Despite both Marsh and Cope being financially secure and able to fund their own extensive investigations, in due time they would drive each other into financial ruin, all over bones.

Initially, these two leading American paleontologists collaborated and even named newly discovered fossilized species after one another. Animosity simmered, however, when Cope discovered that Marsh had paid some of Cope's assistants to send recovered fossils to Yale rather than to Cope's home in Philadelphia. In 1870, ani-

*Othneil Marsh (a) and Edward Cope (b) were prominent paleontologists in the late nineteenth century. Their animosity toward each other was legendary and extended to sabotage and other forms of chicanery. (c) Marsh (standing, center) resorted to arming his assistants, at least for this portrait.* (A) MATHEW BRADY/LEVIN CORBIN HANDY; (B) FREDERICK GUTEKUNST; (C) JOHN OSTROM/PEABODY MUSEUM.

mosity boiled. Cope published a description of a new marine reptile with the head mistakenly drawn on its tail end. Marsh delighted in pointing out the error. Cope tried to buy all the copies of the publication to limit the damage before he could correct his mistake.

At this time, the fossil fields of eastern Colorado, Wyoming, Kansas, Nebraska, and Dakota Territory were ripe for discovery. Both paleontologists hired teams of diggers to excavate bones and ship them east for study and classification. As a result, Cope described 56 new species of dinosaurs and extinct mammals. Marsh described 80. Each tried to keep his most productive digging fields secret from the other, but the competing crews contained trespassers, spies, and double agents, who would keep their employer posted yet at times leak information or ship recovered fossils to "the enemy." The presence of nearby hostile Native Americans only heightened the drama of faking identities, thieving, rock throwing, dynamiting each other's dig sites, and gun drawing. The Bone Wars were on.

Marsh and Cope made multiple accusations back and forth, at first solely within scientific circles, where their mutual hatred was well known; but in due time, the vitriol spread to front-page newspaper reports of plagiarism, financial malfeasance, and scientific skullduggery. On death, Cope donated his skull to science so that the size of his brain could be measured; this was a trendy thing to do at the time. He hoped that Marsh would do the same and that Cope's brain would prove to be larger. Marsh resisted. Regardless of brain capacity, their egos were beyond measure.

But what a legacy they left, though in their haste not every classification they made has stood the test of time. When they began work, only 18 dinosaur species were known in North America. Between them they described over 130 more. These famously include Marsh's original descriptions of *Triceratops* and *Stegosaurus*. Both men proudly named multiple species after themselves, and others did so out of respect for one or the other. Together they brought many tons of specimens east for study and display. Marsh's collections reside at the Smithsonian and Yale Peabody Museums, Cope's at the Academy of Natural Sciences in Philadelphia. Despite their mutual hatred, Cope and Marsh changed forever the way natural history museums are conceived, built, and valued. Who won the Bone Wars? We did.

From Buckland's first description of a dinosaur in 1822, fossil hunting has expanded broadly from its primarily North American and European beginnings. Examination of rock exposures in South America, Africa, Greenland, Pakistan, Antarctica, and China have greatly increased the number of extinct species identified and classified. This has led to a far better understanding of the relationship between dinosaurs and birds as well as the path of human evolution, which is the domain of paleoanthropologists.

If some paleontologists were quirky and belligerent, paleoanthropologists seem, by comparison, to be speculative and argumentative—sometimes even verbally duking it out over a single tooth or a fragment of fossilized skull or jawbone. These specialties have also had their eccentrics. Consider Robert Broom, an early twentieth-century physician turned paleoanthropologist. When a patient died, he might bury the body in his garden to have it available for later study. In keeping with a doctor's formality of dress, Broom continued wearing pointed collars and three-piece suits when digging for bones in South Africa. When it got too hot, he would strip naked. It is unclear whether this habit attracted or repelled predators in the area.

Although Broom was eccentric, his behavior and dress were harmless and did not alter the course of science. A contemporary of his was not so innocent, however. In 1908, the discovery of some bone fragments and teeth left the local residents of Piltdown, England, helpless as nationalism, tunnel vision, and wishful thinking hoodwinked the specialists who pondered the finding's significance. These bones certainly have much to teach—but not, as it turns out, about the discovery of a missing link in human evolution.

In the early twentieth century, paleoanthropology was hardly 50 years old. The discovery of fossilized bones that resembled those of humans in the German Neander Valley in 1856 had caused museums to reexamine their own fossil collections. These were bones that had been collected over the previous several decades. In many instances, their fossils matched the recent discovery in Germany, and so they were reclassified as Neanderthals. In due time, scientists determined that Neanderthals were an extinct species and not an intermediary in the development of *Homo sapiens.*

This was an exciting time for the burgeoning field of paleoanthropology. If Neanderthals were not the missing link between small-brained, four-footed apes and humans, what was? Human-like fossils were turning up in France and Germany along with stone tools of a similar age, whereas British investigators were finding tools but no fossils. It was 1912 and Charles Dawson's timing was perfect. As an accomplished amateur scientist, he was privileged to announce his discoveries, which had taken place over the previous several years, to the Geological Society of London—discoveries of fossilized skull

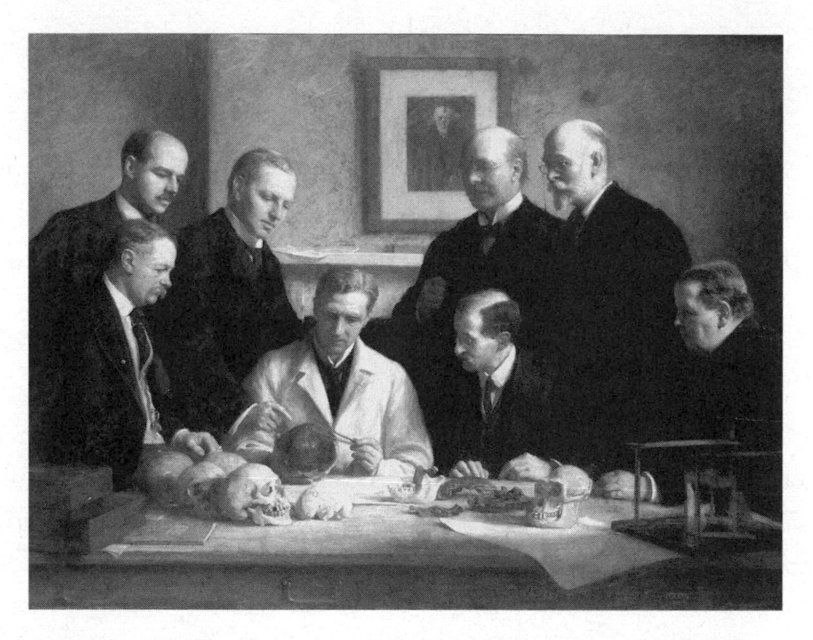

*This 1915 portrait shows osteologists, zoologists, and paleontologists comparing various skulls with the bone fragments of Piltdown Man, which were discovered by Charles Dawson (standing, second from right, in front of a portrait of Charles Darwin).* JOHN COOKE, *PILTDOWN GANG*, 1915.

fragments, a jawbone, and several teeth. They came from a gravel pit near Piltdown, a village about 40 miles south of London. Was this the missing link?

The prevailing opinion was that our direct ancestors were apelike creatures who first developed large brains. Then later, modification in the shapes of the jaw and pelvis allowed this smart animal to manage a varied diet and to walk upright. Civilization eventually ensued. Dawson's discovery fit this opinion. His Piltdown Man (or it could have been a woman) had a large braincase and a primitive apelike jaw with canine teeth that were intermediate between apes and modern humans. It had the necessary characteristics to be the missing link.

The Piltdown fossils became the pivot point for any evolutionary theory that followed, usually using Dawson's findings for support, but at least having to address them if there was disagreement. The discovery stirred national pride in both paleoanthropologists and the British public. Museum exhibits flourished, yet casts of the fossils attracted only a fraction of the attention given to drawings and models that imagined the living appearance of Mr. Piltdown. He became part of popular culture and was the subject of numerous newspaper articles, letters to the editor, postcards, books, and monographs. Mr. (or Ms.) Piltdown was a rock star.

Skepticism about the fossils' authenticity came early and in several forms. The fragmented and limited skeletal pieces were, perhaps conveniently, missing their most diagnostic portions. Was the gravel bed where Dawson found the fossils as ancient as he indicated, or were the fossils from a more recent era? Were the jaw and cranial fragments from the same species? From the same individual?

To seek answers, many investigators had to resign themselves to study casts of the fossils rather than the originals. In 1915, a Smithsonian scientist complained about the inadequacy of studying the casts. Even so, he remarked that the skull fragments and jaw were too different from each other to assume that they were from the same creature. He felt that the skull came from a human, not old, and the jaw from a species of ancient chimpanzee.

Discovery of other human-like fossils in China and Africa in the decades that followed produced conflicting information, and general agreement on how it all fit together was lacking. During this time, Dawson's reputation and his findings' major significance remained intact, although he died in 1916.

The dogma generally prevailed until mid-century, when investigators subjected the bones to a new test: fluorine analysis. Fluorine seeps into buried bones from surrounding groundwater, so fossil bones from the same animal and resting next to each other will have equal fluorine content. The scientists did not undertake the investigation lightly, since they had to destroy small portions of Mr. Piltdown in the process. The results showed that the skull and jaw fragments contained different levels of fluorine. Additional analyses indicated that the bones were not nearly old enough to be any missing link. Furthermore, the skull was human and the other bones were not. Under microscopic scrutiny, the teeth showed evidence of having been filed down to disguise their origin.

If confirmatory evidence was needed, it came in 2009. Both computed tomographic scanning and DNA analysis indicated that the teeth and jaw were from an orangutan. The CT scan also revealed an off-white putty covering the bones' surfaces and sealing interior voids, which were filled with grains of sand. It is likely that the perpetrator of the hoax weighted the relatively modern bones with sand to give them the heft an expert would expect from a fossil. Finally, the fraudster had stained all the surfaces brown to give the bones an ancient, homogeneous appearance.

Who was the perpetrator? Nobody knows for sure. Conspiracy theories abound. The most likely suspect, of course, is Charles Dawson. He was an accomplished amateur geologist and archaeologist with knowledge and experience about how ancient artifacts looked. It turned out that he was responsible for several small-time antiquarian fakes. Dawson also pined for acceptance within the British scientific community and made persistent but futile attempts to join the Royal Society. In addition, he yearned for knighthood, an honor that bypassed him because of his early death but which was bestowed on several of his contemporaries who studied the Piltdown fossils.

How did this happen and what can we learn? First, as con artists do so well, this fraudster showed his audience what it wanted to see. The experts suspended critical judgment and discounted the red flags in their zeal to accept a national treasure, one that put Britain at the forefront of scientific discovery. It proved to be an embarrassment to the objectivity of science, albeit one that eventually corrected itself. The hoax would have been uncovered much earlier had more investigators been able to examine the actual fossils rather than replicas. (By contrast, in every scientific discipline today, some people are calling for the original data to be made available for all to scrutinize.) Fluorine, DNA, and CT analyses eventually exposed the fraud.

The lesson that the Piltdown bones teach is far more relevant and timeless than any insights that would have been gained had the discovery been real. It is human nature to see what we want to see, especially when it conforms to our preconceived notions. Instead, we should take the evidence for what it is and even then retest its validity from time to time, especially as new measurement technologies come along.

What measurement methods are currently in use? When you read that Lucy lived 3.2 million years ago and that Mungo Lady lived 42,000 years ago, you may wonder how scientists can know for certain. Since the 1950s, when fluorine analysis was applied to Mr. Piltdown, scientists have devised a number of clever and sophisticated tests, described as either *relative* or *absolute* dating, to determine the age of ancient bones.

Fluorine analysis is an example of relative dating. It cannot determine how old a bone is, just whether it is the same age as the one discovered next to it. Another technique is stratigraphy—guilt by association. Deeper layers of rock are generally older than those above them. If bones are contained in a sediment layer of known age, then the bones can be assumed to be equally old. Bones in more superficial layers will have been more recently deposited.

Absolute dating is far more precise and requires an understanding of nuclear physics, which is not my forte, but here is the gist. Bodies and bones decay, as do a number of chemical elements. Blood alcohol levels do the same once the party is over. Consider the aftermath of tossing back four shots of tequila in rapid sequence. Within a short time, I would be seriously drunk, and then my liver would slowly metabolize the alcohol. By the next morning I might not be feeling too well but could pass a breathalyzer test. A week later I would be feeling fine, no alcohol left in my system, but an assay of my liver enzymes might hint at the indiscretion.

Similarly, every living organism, by eating, drinking, and breathing, takes up small quantities of radioactive elements in proportion to each chemical's concentration in the environment. Uptake stops at death, and the embedded radioactive elements gradually disappear as they convert to stable forms. If an assay shows that the concentration of the element's radioactive form is nearly equal to that found in the environment, then death was recent. If none of the radioactive form persists, then the bones are old. Intermediate levels equal intermediate ages.

Wood or charcoal and grave goods of biological origin (for example, leather, fabric) found near the bones can be similarly assayed for their level of radioactivity in order to validate the age of the bones, as can analysis of the surrounding soil or rocks. Used in combination with other impressively named tests, including luminescence, electron spin resonance, paleomagnetism, and fission tracking, fossil dating is reasonably precise today. These tests would have exposed the Piltdown hoax in an instant.

What works for identifying the age of bones using radioisotopes also works for determining whether an animal remained in the same area throughout its life or moved from region to region. Strontium occurs naturally in two forms, and the proportion of one to the other varies from one geographical region to the next. These forms are in the soil and find their way into bone via the food chain. Both forms of strontium are taken up and fixed forever in developing tooth enamel, so analysis of the ratio indicates where on earth the owner spent his youth. In bone, by contrast, previously acquired strontium atoms are slowly replaced by ones more recently acquired, so skeletal analysis identifies where the owners spent the last 5 to 10 years of their lives. This test has proved useful to study migration patterns and to ascertain the homelands of Bronze Age European warriors. If the warriors came from different areas, they were trained warriors rather than local villagers, which indicates a turning point in the history of warfare.

In recent years, investigators have successfully isolated DNA from both bone and tooth fossils as old as several hundred thousand years. The good news is that only minuscule quantities are needed for analysis, and the findings have at times been surprising. Who would have previously guessed that Neanderthals and humans interbred with some regularity? The bad news is that since only minuscule quantities of DNA are needed for such analyses, the samples are subject to contamination. At times what was thought to be fossil DNA turned out to be modern DNA from somebody handling the bone or from hitchhiking microorganisms.

Another example of what chemical analysis can teach about skeletal remains involves alcohol. Tetracycline is an antibiotic that coincidentally has an affinity for bone, and Nubian mummies from almost 2,000 years ago are laced with it. The likely source was beer, which they made from grain contaminated with a bacterial strain that produced tetracycline. It is not known whether the Nubians used the beer as an antibiotic or that the antibacterial properties of their libation were side benefits.

Enough for physics and chemistry. What types of information can anthropologists derive by gross physical examination and measurement of bones? Plenty, both from the bones themselves and from the location where they were found. Careful examination can determine species, sex, body habitus, age, health and nutritional status, and injuries, both new and old. Hack marks made with stone tools indicate butchering practices and vary according to how fresh the carcass was when butchered. Certain patterns of hack marks on human bones likely indicate cannibalism. Skeletons excavated from mass graves with limbs askew and containing fresh fractures imply genocide. Grave goods (for example, pottery, weapons, and jewelry) placed to accompany the deceased into the afterworld give indication to that culture's beliefs and the family's economic status. Unintentional grave goods, including pollen and insect exoskeletons, can help scientists determine the season of death and the climate at the time.

By comparing skeletal elements over time, anthropologists can also recognize changes in human habits. For example, after our ancestors controlled fire and began cooking their food, human jaws became thinner and weaker, because cooked meat and roots were easier to chew. Likewise, human thighbones are now thinner than they were just several hundred years ago, likely an activity-related finding.

Overall, I think that anthropologists have a raw deal. They have to confront cannibalism, interspecies breeding, and genocide. Based on what I see on *Nova* and in *National Geographic*, anthropologists spend a sizable portion of their lives squatting, sitting cross-legged, or lying on their side in a squared-off hole somewhere outside Timbuktu while tediously extricating each fragmented treasure with dental picks and small brushes. It may also be frustrating. In one instance, local diggers were intentionally fragmenting relics to increase their per-piece payment. For all these reasons, I am not surprised that a publicity-starved discoverer or an overzealous reporter might tend to sensationalize a find. I was skeptical when I read that the cast of a boy found in the baths of Pompeii indicated that he was there looking for his parents at the cataclysmic moment. How sweet. Maybe he stepped inside to go to the bathroom; but this scenario is not one likely to attract media interviews and funds for continued research. There are multiple instances where the remains of two people have been found in the same grave, apparently embracing. How dear. The news media goes wild, but nobody really knows how these two skeletons ended up so positioned, and less newsworthy theories get no attention. Overall, such imaginative hoopla is entertaining and relatively harmless.

Not so in the late nineteenth century, when bone studies moved far beyond the harmless art of phrenology—determining one's character and intelligence by palpating one's cranial prominences. Rather, anthropologists (at the time, all white men) went far astray and set out to make facts fit their pet theory that Caucasians were superior to other races. These "scientists" collected skulls by the tens of thousands, measured cranial capacities, and used the results as markers for brain size and cognitive superiority. This led pathologist and anthropologist Paul Broca to conclude, "In general, the brain is larger in men than in women, in eminent men than in men of mediocre talent, in superior races than in inferior races. Other things being equal, there is a remarkable relationship between the development of intelligence and the volume of the brain." How wrong he was, but in the meantime and equally divided between the United States and Europe, museums collectively came to house probably a million sets of Native American remains along with smaller but impressive numbers of Caucasians', African Americans', and indigenous peoples' remains worldwide. The museums in major cities, fanned by major egos, competed with each other to have the largest "bone room," ethical considerations be damned. This embarrassing frenzy died out before World War II, and museums that kept their bone rooms did so in the context of studying human origin and evolution rather than racial inequality.

This racially motivated fiasco eventually resulted in the passage of the Native American Graves Protection and Repatriation Act, which became US law in 1990. It requires institutions and agencies that receive federal funding to return human remains as well as funerary and sacred objects to the Native American descendants of the original owners.

Rampant pillage of gravesites has been further quelled by other regulations. All proposed archaeological excavations now have to be approved in advance by the state's historic preservation office and by the Native American tribes who are traditionally from that territory. Approvals typically stipulate that if human remains are encountered, the excavation stops. Either the tribal archaeologist, the historic preservation archaeologist, or the excavating archaeologist will remove the human remains. Part of that decision is based on whether or not a contractor is waiting to continue a construction project. The tribe usually receives the human remains directly without any scientific analysis. That may seem unscientific; but the remains are somebody's relative, and what right do I have to put these emotionally laden materials in a laboratory drawer or museum display case?

I like the compromise that the indigenous people and the anthropologists agreed to in Australia. Remember the Mungo Lady? Her cremated remains are under double lock at Mungo National Park. The local indigenous people control one key, the archaeologists have the other. Both are required to open the vault.

*Chapter 13*

# THE BUSINESS OF BONES

Over the centuries, bone has provided tools for a spectrum of businesses, including architecture and drafting, carpentry, sail making, rope making, bookbinding, and pin making. Furthermore, because of bone's complex composition and enduring structure, it continues to support myriad business endeavors as a material. Rather than offer an encyclopedic list of bone's contributions to paint, soap, and sugar manufacture among many others, I'd like to highlight, in roughly chronological order, eight astoundingly diverse ways that enterprising individuals have commercialized bone.

The first is bone's role in revolutionizing the fashion industry. Until buttons came along, clothing was loosely draped over the torso. Sadly, the healthy human form was all but invisible, and this was especially so among the upper class because extensive drapery denoted wealth. Yet it was a problem to prevent these flowing garments from sliding off—that alluring moment often captured by sculptors of classic beauties.

The initial answer was long bronze or bone pins that pierced the fabric and stabilized the folds. At first buttons were only ornamental; and when they initially became functional, they were secured through loops of cord passing along the edge of the garment. The reinforced buttonhole came along in the thirteenth century.

Now garments could be formfitting. The more buttons, the closer the fit. Detachable sleeves fixed with buttons became the vogue. This allowed for mixing and matching wardrobe items and for selective cleaning.

The wealthy exhibited vast arrays of ornate glass and metal buttons as status symbols, far beyond functional requirements. The epitome of button excess is likely an outfit worn by the king of France in 1520. It had 13,600 buttons, each with its own buttonhole.

The lower classes naturally parroted this lavish display of buttons, but their buttons were made of inexpensive bone. Some were undoubtedly homemade, but industries sprung up, sufficiently so that French button makers formed a guild in 1250. In Konstanz, Germany, archaeologists have recovered 300,000 perforated strips of cow bone left over from a button- and bead-making industry that thrived there between the thirteenth and sixteenth centuries. As a business boost, the Holy Roman Empire spread its influence during this time, and the demand for rosaries increased. Even impoverished believers could afford rosary beads made of bone.

The wealthy also had their economies. They fastened their undergarments with bone buttons and saved the fancy ones for all to see. Gentlemen generally fastened their own buttons. Because a lady's garment might be almost entirely covered with buttons and buttonholes, the wearer had her aide or aides perform the tedious task of buttoning and unbuttoning. It was at this time that buttons shifted sides on women's clothing, which eased this task for predominantly right-handed dressers.

Time flies. The fashionistas die. Centuries pass. The clothes, fancy and plain, disintegrate. Yet the buttons survive. Recovered from archaeological sites, these humble disks chronicle bygone fashion and material culture. Nylon zippers and Velcro hook-loop fasteners, noisy as they are, may prove to be equally durable and equally valuable to historians, but humble bone buttons over a far longer period continue to quietly reflect human culture.

It is not clear whether any of the Dark Age button manufacturers themselves became wealthy, but somebody was getting rich, because a large market developed for small "caskets"—actually jewelry boxes—that were frequently given as engagement gifts. Herein the betrothed could keep her gems, love letters, and other precious items.

An enterprising Florentine merchant and diplomat named Baldassare degli Embriachi seized the marketing opportunity. He began manufacturing hexagonal and rectangular boxes ornately decorated with bone carvings to satisfy the luxurious tastes of European royalty and nobility, who had a desire for ivory but not necessarily the means to afford it. (Elephants sighed with relief—momentarily.)

Embriachi's craftsmen carved low-relief images on rectangular segments of bone, usually from horses or oxen. The carved figures often depicted tales from mythology, medieval romance, or the Bible. On the boxes, the artisans surrounded the carvings with elaborate frames inlaid with wood, horn, and bone. The workshop also produced family altarpieces, and prestigious donors commissioned a

*Ornately carved panels of bone embellish this box, which is one of many turned out by Baldassare degli Embriachi's workshop around 1400.* LOS ANGELES COUNTY MUSEUM OF ART.

few magnificent ones for monasteries. The workmanship is astounding in detail and volume, especially considering that it was all done with hand tools.

Embriachi eventually moved his workshop to Venice and turned the enterprise over to his two sons. Production spanned nearly 60 years, centered on the year 1400. Judging by the number of pieces that I have seen in museums, on Google Images, and on Pinterest, the Embriachi enterprise must have produced many hundreds, if not thousands, of these luxury goods.

For several reasons, products from the Embriachi workshop cannot be specifically dated or even clearly attributed to Embriachi versus one of his competitors. This conundrum for art historians and collectors stems from the fact that multiple artisans contributed to the production of any single object, making it difficult or impossible to confirm a signature style or evolution of style. Furthermore, to meet the pocketbooks of Embriachi's clientele, the workshop produced secular objects in various qualities throughout its existence, again confounding any attempt to note a refinement of motif or detail over time. Many fine art and decorative art museums display examples of these alterpieces and boxes. The caskets, along with chess and backgammon boards, appear from time to time in high-end art auctions. Prices range from thousands to tens of thousands of dollars. If the new owner has any cash left over, I guess they can stash it inside their bone box.

Along with Embriachi caskets, bone china is well represented in museums and also in refined households around the world. *Refined*, by definition, is the removal of unnecessary elements, leading to elegance in appearance, taste, or manner. For china, that would include reducing the thickness of plates, saucers, and cups as much as possible. But in so doing, the tableware becomes fragile. It would be shameful to serve afternoon tea in chipped cups or have a weighty roast pig break a platter. In 1797, Englishman Josiah Spode II came to the rescue. He perfected a porcelain-manufacturing process that his deceased father and others before him had experimented with. Josiah II's main ingredient was bone ash, which is the calcium and phosphorus compounds left after bone is baked in a high-temperature, oxygen-starving oven. Twelve parts of bone ash, eight parts of china stone (a granite-type mineral), and seven parts of china clay (an aluminum and silicon-containing mineral) were, and still are, the formula for bone china.

Spode's formula made his china the hardest of all porcelain ceramics, which means that the tableware can be delicately thin yet quite durable and chip resistant. An added benefit is its translucency, which adds to its refinement. What a great gig for lowly bones salvaged from slaughterhouses.

At the same time Spode was baking, Napoleon was fighting; and before he was defeated in 1815, as many as 100,000 French prisoners of war whiled away time in British custody under minimum-security conditions, some for as long as 10 years. Prior to conscription by Napoleon, many of the French soldiers had been furniture makers, metalsmiths, and weavers. To pass the time and at the encouragement of their captors, these craftsmen began salvaging cooked mutton bones from the prison kitchen, cleaning and bleaching them, and then building intricate models, which they sold or traded for fresh produce in nearby towns on market day. The output included ship replicas made principally of bone along with bits of gold and silver foil, silk, and turtle shell provided by the locals. The ships were mostly idealized versions of contemporary British naval vessels, since the modelers had no working drawings. Instead they designed from memory and imagination.

At times the models included moving parts, such as retractable cannons. They are breathtakingly beautiful and are highly sought-after collector's items, often going for tens of thousands of dollars at auction. The US Naval Academy has the largest collection. The Channel Islands Maritime Museum in Oxnard, California, also has a collection worth visiting.

The French prisoners also made ornately crafted dollhouse furniture complete with sets of china (bone china, presumably), game

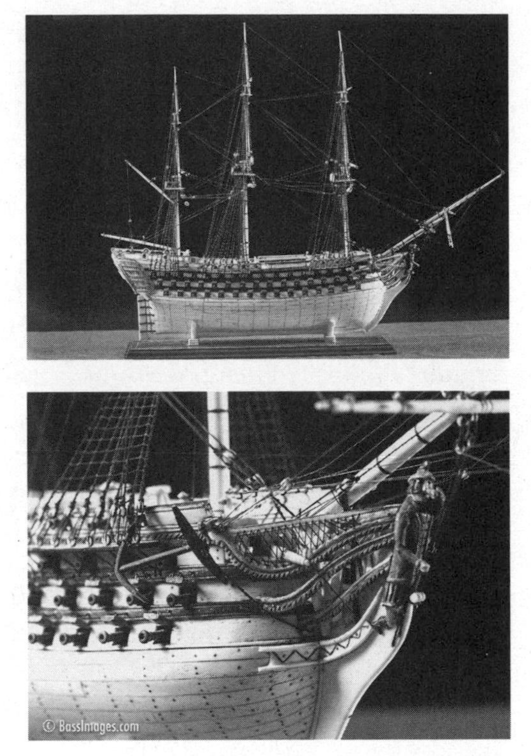


*During the Napoleonic Wars in the early 1800s, French soldiers imprisoned in England turned their crafting skills into a cottage industry by making elaborate models from stewpot bones and selling them to local British townspeople. The ships were designed from memory and imagination and contained many moving parts, including retractable cannons.* CHANNEL ISLANDS MARITIME MUSEUM. © BASS IMAGES, THOUSAND OAKS, CA.

boxes and dominoes, and crank-driven spinning wheels with dancing performers. Two ironies are apparent. Among their crafts were working replicas of guillotines, which may have been the sailors' fate had they returned home before Napoleon was defeated. Also, at the end of the war, a number of prisoners chose to stay on and continue their modestly lucrative bone-crafting cottage industry rather than return to France and to an uncertain future.

⯁⯁⯁⯁

JUMPING FORWARD 50 YEARS and across the Atlantic, what do 30 million American bison, the completion of the transcontinental railroad, and the discovery of superphosphate fertilizer have in common? A lot, as it turns out. In 1868, they became critical elements in the formation of an industry, one that thrived for 20 years. It helped finance the settlement of the Great Plains, ensured solvency of numerous new railroad lines servicing the settlers, and provided vital fertilizer for crops across the entire continent.

This unlikely confluence starts with phosphorus. Without knowing why, early humans discovered that their crops thrived when they planted seeds into soil containing ground-up bone. In 1840, the answer became clear. Phosphate, the P in the NPK that is listed on every bag of fertilizer at the garden center, is crucial for robust flowering, fruiting, and root growth. Bone is an excellent source of phosphate because the chemical composition of hydroxyapatite, the compound that makes bone strong and hard, is $Ca_5(PO_4)_3(OH)$. This indicates that there are three phosphorus atoms in every molecule. In this form, however, the phosphate is not easily dissolved, meaning it takes a long time for a plant to incorporate it.

A few years later an enterprising chemist mixed bonemeal with sulfuric acid. This changed the phosphate into a form—superphosphate—that plants could readily access. They loved it. Farmers couldn't get enough of it.

During the same years, pioneers were rambling west across the Great Plains with railroads rumbling not far behind them. The Native Americans and the roaming buffalo both proved vexatious to the western migration, and it became government policy to exterminate the bison as a means of subduing the Indians. Furthermore, the bison herds were collision hazards for locomotives, which could not stop quickly. To escape blizzard winds, the bison particularly liked to stand on the tracks where the line cut through the hills. Trains could be halted for days. For these reasons, it became common practice for hired marksmen to shoot the bison from moving trains. Their hides might be harvested. The remains rotted in the sun. Over roughly 30 years, tens of millions of bison were reduced to a few thousand.

Bison bones covered the prairie. Wherever a railroad line passed nearby, it became feasible, actually lucrative, to pick these bones up, load them on a train headed back to St. Louis, Detroit, or Chicago, and sell them to fertilizer plants. Baked and ground, bone also found purpose for filtering sugar during the refining process.

The homesteaders benefited, particularly in their first year on the prairie, when they did not have crops to trade for needed farm equipment and food staples. The railroads benefited because the trains brought consumer goods west and would otherwise return empty were it not for the bones.

An industry blossomed. A homesteading family or team of itinerant bone pickers could harvest a ton a day, cart the bones into town, and sell them at the local railhead for $5 to $8. This was at a time when a family could manage fine on $10 a week. Brokers sprung up in every town along the railroad line. They bought bones from the pickers, stacked them in huge piles, and sent them on their way east with the next passing train. The Native Americans, however, did not participate in this industry out of respect for the bison, which for millennia had provided them food, shelter, and clothing.

As new railroad lines extended west, new opportunities for picking up and delivering bones within practical distances also expanded. Sometimes the bone pickers would anticipate a new railroad extension and wait with their stash until the line arrived. What started in an east–west band across Kansas and Nebraska eventually extended south into Texas and north into the far recesses of Alberta and Saskatchewan, where local boosters eventually arranged to have the name of their town, Pile o'Bones, changed to Regina.

When pickings became slim, bone pickers would throw in antlers, which are also bone, and, according to some accounts, bones raided from Indian graves. The brokers and factory owners didn't seem to mind. Itinerant teams would also burn the prairie to find bones otherwise hidden in the tall grass. They cannily started the fires close to the railroad tracks so that they could plausibly attribute the flames to sparks from locomotives.

Accounts of the tonnage of the bones collected are mind-boggling. Although nobody kept an overall tally, several observations are revealing. Brokers in Minot, North Dakota, moved 375

*Over several decades in the late 1800s, roughly two million tons of bison bones were collected from American prairies and shipped east to be ground into fertilizer.* Burton Historical Collection, Detroit Public Library.

tons of bone in 1887 and again in 1888. This grew to 2,200 tons in 1890. By mid-June of the same year, pickers had brought the remains of over 100,000 animals into Saskatoon alone. By August, a shortage of railroad cars resulted in 165,000 skeletons stacked in one pile waiting transport. Other smaller piles rested nearby.

During its existence, bone picking was roughly a $40 million industry involving 2 million tons of bones—enough to fill two rows of boxcars crossing the continent from San Francisco to New York. By comparison, the previous "big" industries of bone buttons and jewelry boxes were eclipsed many times over. Yet none of these industries were sustainable. By the early 1890s, railroads had expanded into all of the areas previously occupied by the bison, the prairie had been picked clean, and the industry collapsed. Fertilizer manufacturers turned to a mineral form of phosphorus and continued making fertilizer.

Now derived from the meatpacking industry, you can buy bonemeal at the garden shop. It provides much-appreciated phosphorus nutrition for your plants. Let a steer eat your flowers, butcher the bovine, grind its bones, fertilize your plants again. The phosphorus goes round and round, but bison are forever out of the loop.

A second bone industry dawned on the Great Plains about the same time that bone picking picked up. Little did Joseph Sherburne and Ponca Chief White Eagle know that trading in corncob pipes would set off a fashion frenzy that would last 30 years. Sherburne had obtained a license to trade with this group of Native Americans, who in 1878 were living in Indian Territory, now Eastern Oklahoma. Among his trade goods were pipes fitted with bone stems. They quickly sold but without comment from the purchasers. When Sherburne next visited, Chief White Eagle showed him an elaborate neck ornament made from bone pipe stems strung on buckskin strands. White Eagle requested more pipe stems—in quantity.

For centuries preceding, Native Americans had favored necklaces made with long, slender tubular beads. Gravesite excavations have found such beads made from bird bone, conch shell, and rolled copper. The Native Americans wore these beads around their necks and in their hair, which may account for their name—hair pipes—but nobody knows for sure. The tribes, however, clearly valued them, because one early European trader noted that one conch shell hair pipe, about the length and girth of a man's index finger, was worth four deer skins. Lewis and Clark took a supply of hair pipes with them on their Voyage of Discovery and would disperse one or two per tribal chief according to each one's eminence.

Recognizing the Native Americans' fascination with hair pipes and their commercial value, government traders offered silver hair pipes beginning in the early 1800s. Private traders preferred less expensive hair pipes made from large West Indian conch shells.

In New Jersey, a cottage industry experienced in converting shells to wampum (beads) began making hair pipes in industrial quantities. By 1830, a family of Campbells was so successful in doing so that they paid their neighbors to perform the preparatory steps, and then they hand-lathed, drilled, and polished the beads. Drilling a narrow hole the entire length of the 4-inch-long hair pipe was a technical challenge, and the family kept their method secret. Over generations, the Campbells perfected machines such that eventually a worker could produce 400 pieces in a day.

Traders of every ilk then distributed them to Plains Indians, mostly men, who used them as hair adornments and ear pendants. Tribes east of the Mississippi did not seem to be interested (maybe jaded by New York fashion), and the style spread slowly from the Great Plains to the Rockies and farther west.

The conch shell hair pipes broke easily, and photographs show wearers styling damaged goods. Along came Chief White Eagle and his unusual request for hundreds of corncob pipes. Trader Sherburne contacted a wholesaler in New York from whom he had previously bought glass beads and asked him to mass-produce bone hair pipes, which he did. The Armour beef-packing plant in Chicago sent bovine leg bones to New York, where they became hair pipes and got sent west.

The bone beads mirrored the size and shape of the conch shell hair pipes and were more durable. They were also considerably cheaper, selling for 10 to 15 cents each, depending on their length, compared with 50 cents for those made from shell. Within 10 years, the 60-year-old Campbell business for conch shell beads folded.

Bone hair pipes became widely available in the 1880s, which was a time of economic and mental depression for the Plains Indians. The buffalo were gone, reservation life was strange and tedious, and government rations were meager. Bone beads were plentiful and cheap, and elaborate hair pipe adornments symbolized prosperity. Even though conch shells were whiter and did not have the dark streaks characteristic of bone, the price and durability were right, and the Native Americans captured the opportunity to regain a bit of dignity.

Not only did they wear hair pipes during their own ceremonies,

*Bone beads manufactured in New York became highly desirable decorative items among the Plains Indians in the late 1800s.* LIBRARY OF CONGRESS.

they also did when socializing with other tribes, visiting the Great White Father in Washington, and participating in or attending Buffalo Bill Cody's Wild West Show. Photographs from the time show elaborate choker-type necklaces and bandoliers of hair pipes in long strings and wide bands. The pièce de résistance, however, was the breastplate—numerous hair pipes strung on buckskin and arrayed horizontally on two or more vertical rows.

The trend likely started with the Comanche tribe in the mid-1800s and spread quickly through the Great Plains. When the durable, cheap osseous version became available, it seemed to create a contest of "my breastplate is bigger than yours." The record is likely 140 hair pipes arrayed in two rows. Workmanship also counted, and in the early 1900s, a well-made breastplate had the same value as a horse.

Then the hair pipe trade dried up. Buffalo Bill's death in 1917 and the closure of his Wild West Show, which employed 65 Sioux performers, may have contributed. Also, by that time the Native Americans had a large number of entirely durable hair pipe adornments on hand, so they stopped buying the beads. Hence, the traders ceased stocking them.

The story of hair pipes is one of cooperation. Two individuals, White Eagle and Sherburne, and two extended families, the Plains Indians and the Campbells, had something that the other wanted, which is ironic considering the historical animosity between the two cultures. The cooperation turned out to be a win-win. The story is also one of creative destruction. Bone supplanted shell as the raw material, and industrial-scale manufacturing squelched a successful family business. Then the new product saturated the market and disappeared. Sound familiar? The next business, however, seems to be enduring well.

Since bone decays far more slowly than other human tissues, obtaining burial space in densely populated areas eventually becomes problematic. Historically in European cultures, individuals of high standing were buried within the church, others nearby outside. Families could only rent plots, however, sometimes for as little as 20 years. After that, making space in the boneyard for the newly deceased necessitated removal of old skeletons. Workers exhumed, sorted, stacked, and stored the old bones compactly in underground crypts or catacombs.

The largest catacomb by far is in under Paris, which contains the bones of an estimated 6 million bodies in an extensive network of tunnels and chambers. Much of the city is built from the limestone blocks removed from that subterranean quarry, whose passageways and rooms remained empty until the late 1700s, the time when cemetery storage of old bones rose to crisis level.

Once the plans were made, it took over two years for nightly processions of wagons to discretely deliver the bony remains from Paris's cemeteries. This included at least 2 million skeletons from a single cemetery that had been in use for 600 years. Along the quarry's corridors, the workers stacked thighbones and skulls into retaining walls, behind which the other bones were unceremoniously tossed. To add a semblance of order, the workers placed marble plaques identifying the cemetery from which any given jumble of bones had originated.

Over the following centuries, the Paris catacombs have proved to be an enduring attraction. At first curiosity seekers, mostly royalty, were allowed to pop down several days a year. Now it is one of Paris's most popular tourist magnets and is open 63 hours a week (closed on Christmas). Buying one of the limited number of advance-purchase tickets online may cut a typical 2-hour wait in half. Once through the door, a staircase spirals down about six stories, and then the dimly lit tunnel meanders for approximately a mile. Calcium carbonate—limestone—forms the floor and low ceiling. The visible walls are calcium phosphate—skeletons.

The bones in the Paris catacombs have little scientific value, since they are jumbled, which makes it impossible to observe any trends regarding an individual's or a group's state of health, nutrition, life span, or cause of death. Rather, the experience is a history lesson—a unique, strange, unforgettable, yet not particularly creepy (at least for me) window on Paris's past.

A visit also provides insight into human nature. Tens of thousands of tourists annually descend into the earth to glimpse their own mortality. Forty-five minutes later, they spiral back up into sunlight. What awaits them? The Paris Catacombs Gift Shop.

The commercial applications of bone described so far valued the dry material either for its mechanical structure and appearance or for its chemical composition. During both world wars, fresh bones were in demand for fabrication of bombs and glue for airplane assembly. For bombs, the key ingredient from bone was fat, from which workers extracted glycerin and turned it into highly temperamental nitroglycerine and then into less irritable dynamite. For glue, the key ingredient from bone was collagen, which they converted into a tenacious adhesive. The source for these bones was Allied citizens, who officials urged to recycle their cooking bones. On one poster a Limey shouts, "I need bones for explosives. Put out your bones for salvage to make explosives, lubricating oil, fire-resisting paint, animal feed, fertilizer, aircraft, camouflage materials, glue, etc." Bone's versatility goes on and on.

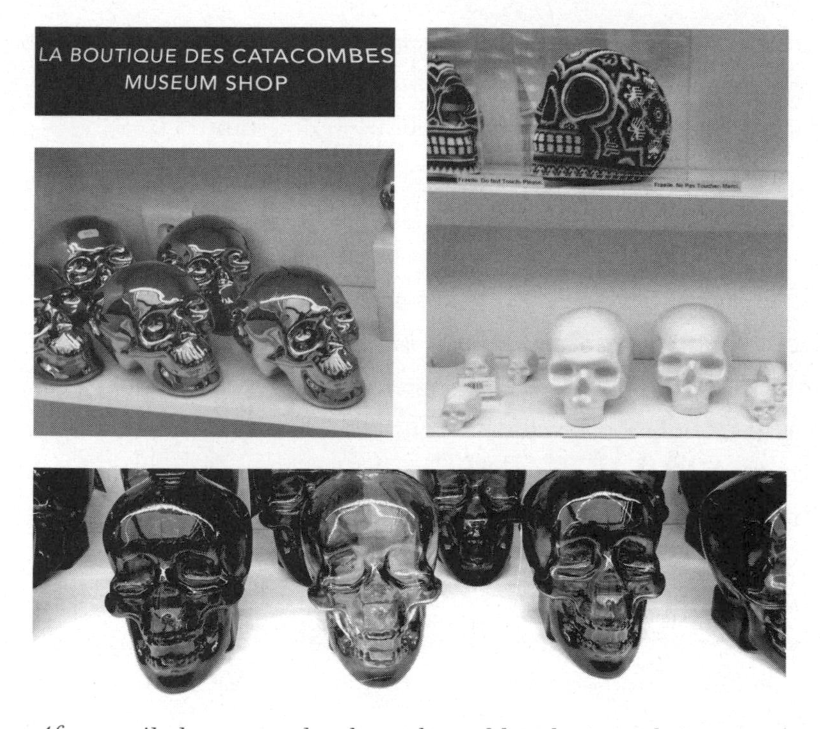


*After a mile-long meander through an old underground stone quarry stacked to the ceiling with bones, a tour of the Paris catacombs brings the visitor back to daylight and street level at the gift shop.* PARIS CATACOMBS.

Modern commercialization of bones extends to their fossilized form. As in many businesses, there are conflicting interests, sometimes monstrous. None are more so than those surrounding the discovery and ownership of Sue, the largest and most complete fossil of a *Tyrannosaurus rex* ever discovered. It started with a flat tire in 1990 in the hinterlands of South Dakota. While the rest of the team from the Black Hills Institute (BHI), the world's largest commercial fossil dealer, went to town to get the flat fixed, experienced amateur paleontologist Sue Hendrickson decided to have a look at a previously unexplored cliff. Along its base, she found several small fossilized bones. Looking up, she spotted the end a large fossil sticking out. On return the team, led by BHI owner Peter Larson, recognized her find as a *Tyrannosaurus rex*, the top-of-the-food-chain carnivore from the late dinosaur era 66 million years ago. In honor of the discoverer and his then-girlfriend, Larson named the beast Sue, although the sex of this or any other dinosaur has never been determined.

On a handshake deal, Larson paid landowner Maurice Williams $5,000 to extract the entire find, which proved to be 90 percent complete and therefore by far the largest and best-preserved *T. rex* skeleton ever discovered. Larson removed the fossils to the BHI lab in Hill City, South Dakota, for cleaning and eventual display or sale. Before either could happen, the FBI swept in and confiscated not only Sue but Larson's entire fossil collection and business records. The Feds were not so much interested in Sue as in other BHI fossils that they alleged Larson had taken from public land and then concealed their provenance to facilitate overseas sales. But ownership of Sue was also in dispute between Larson, Williams, and the federal government, to whom Williams had leased his land. In the end, the court sent Larson to prison for two years and awarded ownership of Sue to Williams.

Williams then contacted Sotheby's to sell Sue. At the auction in 1997, some private collectors competed with several natural history museums for ownership, yet all but one of each were priced out as the bids passed an astounding $7 million mark within 10 minutes. The representatives for Chicago's Field Museum of Natural History had reached their predetermined bidding limit, and it looked like Sue would fall into private ownership. On a hunch that one more offer would work, the Field rep bid again. The auctioneer's gavel came down. "Sold to the Field Museum for $7.6 million." Sotheby's got an extra 10 percent for its troubles.

The Field then invested an additional $2 million building a custom support frame for Sue, one that allows removal of any individual bone for study without disturbing the others. And on its opening in 2000, what an impressive display it was. Sue, lunging with mouth open and sharp teeth exposed, greeted museum visitors in the main atrium for 17 years. Surely the Field Museum has recouped its hefty investment with ticket, book, and gift shop sales as well as museum memberships and international visibility as an institution dedicated to research and education. In 2018, the Field Museum moved Sue to her own exhibition hall and replaced her in the main reception atrium with a much larger and older herbivorous dinosaur for the tune of $16.5 million. Sue's replacement will have to work hard to recoup its financial investment.

Coincidentally, publicity surrounding Sue occurred within several years of the debut of *Jurassic Park*, the fanciful, frightening computer-animated movie portraying dinosaurs in modern times. If dinosaur mania needed its flames fanned, this combo did the trick. Fossil hunting intensified as collectors saw million-dollar figures swirling around the discovery and retrieval of Sue-like specimens. Many landowners began selling their property's fossil rights to the highest bidder regardless of the digger's method or intent.

Those interested in fossils' commercial value do not necessarily recognize (or care much) from which layer of rock their treasures are found. They have no incentive to consider the context of their find—for example, what other plant and animal fossils are present in the same layer. By contrast, trained and disciplined paleontologists excavate methodically and respect the fossils' scientific and educational import far above any sense of their commercial value. The pros take pains to carefully record the exact location of any finds in three dimensions and meticulously search the surrounding stratum for other deposits that will aid placing the fossil in context of the world as it then existed.

Furthermore, when a commercially mined fossil goes up for sale, cash-strapped universities and museums typically lose out. If the new owners display their treasures at home, then the scientific and educational value of the fossils plummets, much to the chagrin of professional paleontologists. A writer for *Slate* commented, "There is no more need for self-styled paleontologists than there is for amateur gynecologists." Commercial enterprises counter that the exposed fossils they retrieve would otherwise crumble and weather away, useful to no one. I guess they are saying that an amateur gynecologist is better than none. You decide.

*Chapter 14*

# DOMESTIC BONES

LONG BEFORE BONE WAS PART OF ANY INDUSTRY, early humans were shaping bone at home to enhance their daily lives. For at least 300,000 years, humans have been deliberately working on bone with stone tools for reasons other than defleshing. Variously shaped bones served specific purposes. Sometimes the transformation is obvious. Skull tops became bowls and thighbones became trumpets. Flat objects notecard-size or larger likely came from a whale's jawbone or the shoulder blade of some other large animal. Straws started out as bird wing bones.

The real workhorse for bone crafting, however, does not immediately reveal its origin. It is a bone that humans do not even have but one that our ancestors could easily access—the versatile cannon bone. Horses, cattle, and bison, as well as deer, goats, and sheep, have one in each leg.

In these hooved animals, the cannon bone is actually two elongated metacarpal (forelimb) or metatarsal (hind limb) bones fused together. The cleft on its end suggests the fusion, and an X-ray clarifies it. The cleft separates two spool-like prominences, and no other bone has this distinctive feature.

Although ancient hunters discarded cannon bones because there was little marrow inside and no flesh outside, cannon bones have had many second lives owing to the combination of their straight shape; broad, relatively flat surface; substantial length and thick-

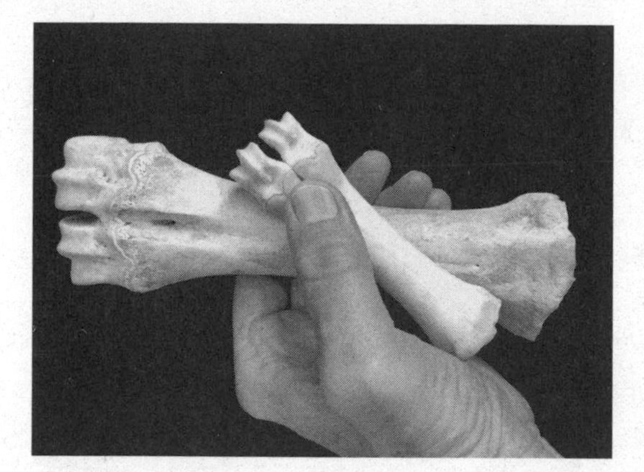

*Hard and dense with only a vestigial central cavity, cannon bones from hooved animals, here a cow and a goat, were the principal osseous material for fabrication of fishhooks, arrow points, domestic utensils, and decorative panels.*

ness; and ubiquity. Enterprising craftspeople from different cultures have shaped them into a panoply of domestic aids and in several instances turned them on end to make pavement—cobblebones.

Along the way, the very first bone tools likely required no modification. Long, heavy bones with knobby ends, particularly thighbones, became clubs. Jawbones served the same purpose. If a club broke or if a long bone was purposely cracked open to slurp the marrow, the sharp-ended remnants readily became daggers. New Guineans extended this art by tapering one end of a cassowary (large flightless bird) or human thighbone and incising elaborate designs. Owners particularly prized the human ones, which came from their father or other respected community member and carried with it the rights and strengths of that person.

Early man quickly realized that daggers and clubs brought him perilously close to his prey, and so he began throwing or shooting pointed sticks from a distance. These arrows, spears, and harpoons (spears with cords attached) added an element of user safety to the hunt. I can imagine, however, that primitive man, after a long stalk and an accurate strike, was entirely chagrined when the pointed stick bounced off his prey or penetrated its skin and then slipped out. Hunters began attaching sharpened points of flint or bone on the end of their sticks. These could more readily penetrate the skin, and when barbs were added, these stayed stuck. The throwing stick,

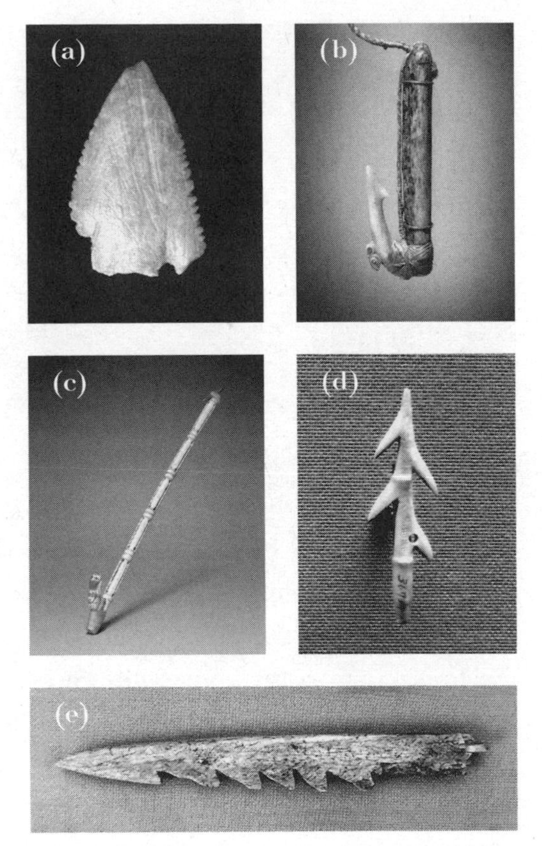


*A multitude of native cultures made barbed bone implements for fishing and hunting before they could make or purchase similarly shaped metal objects: (a) arrow point from Neolithic times in England, about 3000 BC; (b) fishhook, Maori culture, New Zealand; (c) throwing stick for spears, Peru, 200 BC; (d) fishhook, Jomon era, Japan, 2000 to 400 BC; (e) harpoon point, San Juan Island, United States.* (A) PORTABLE ANTIQUITIES SCHEME; (B) MUSÉE DES ARTS PREMIERS IN PARIS; (C) CLEVELAND MUSEUM OF ART; (D) TOKYO NATIONAL MUSEUM; (E) SAN JUAN ISLAND NATIONAL HISTORICAL PARK.

another ingenious invention made partially or completely from bone, effectively lengthened the user's upper limb by almost half and increased the velocity and range of the hurled spear. (Today, plastic ones launch tennis balls for dogs.)

When prey was out of arrow and spear range, the Inuit employed two diabolical methods involving pointed bone. (If you are squeamish, skip this and the next paragraph.) They used the first one to kill wolves and foxes. The hunter started with a narrow, 9-inch-long strip of bone. He pointed, softened, and then folded it on itself three times before wrapping it with a cord. After the bone had dried in this compact form, the hunter replaced the cord with some canine delectable—maybe blubber or fish skin—and left it in harm's way. On discovery, the prey would find it unchewable and swallow it whole. The moisture and heat in its stomach caused the bone to resume its original shape and fatally penetrate vital organs.

The Inuit used a similar method to catch hungry gulls. The hunter attached a short segment of barbed bone to a cord and slipped the assemblage inside a small fish. He then tied the other end of the cord to any convenient, immovable object. An unsuspecting gull would swoop down, take the bait, and attempt to fly away. The barb would turn in its throat and promote its demise.

To diversify their diet, indigenous people in all climates caught fish. Barbs again proved their worth, not only on spearheads but also on fishhooks made entirely of bone or from bone lashed to wood.

Birds also fell victim to another bone-made weapon—bolas, which consisted of several large bone beads threaded onto interconnected cords. When thrown, they would entangle legs and prevent escape. Bones were perfect for this purpose because hunters could drill holes through them. Stone was too dense to drill, and wood was too light to throw with force.

For the same reasons, fishermen made net sinkers from bone, and they used bone shuttles and gauges to make fish netting. Arctic dwellers also exhibited remarkable ingenuity in crafting fittings from bone for their kayaks and sled dog harnesses. Making buckles and toggles not only was expedient but may have provided distraction during long, dark winters. Some of these harness accessories had moving parts and are conceptually identical to ones found on dog leashes and backpacks today. Plains Indians straightened their arrows with "wrenches," which were flat pieces of bone with central holes. A hunter would heat an arrow's wooden shaft and then insert it into the wrench to remove any bends. In these instances, bone itself was not responsible for the kill but was an accessory to it.

Bone also played a part in seal hunting, on which the Inuits depended heavily. They used a long, slender feeler, probably made from a whale rib, to probe the location and shape of a seal's breathing hole through the overlying snow. Then the hunter would wait

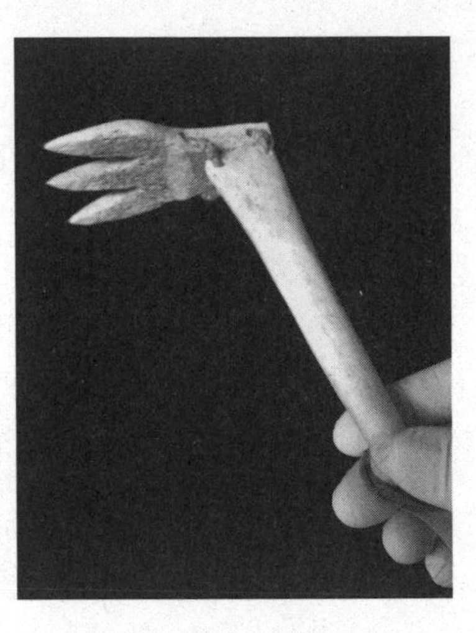

*The Inuits enticed seals to come within harpooning or shooting distance by scratching the ice with bone implements.* NATURAL HISTORY MUSEUM OF LOS ANGELES COUNTY.

patiently. He would scrape the ice now and then with a little hoe-like scratcher made from bone to suggest that another seal was nearby and happily making a new hole. At the right moment, wham goes the harpoon. A successful hunter was likely to open the seal on the spot, eat its liver, and then use "wound pins," made of short, stout, pointed segments of bone, to close the abdominal incision for the trip home. If the seal was too large to load on the kayak, the hunter towed it behind. To keep the catch floating, the hunter made small holes in the skin, blew air into the fat in several places using a bird bone straw, and then sealed the skin openings with conically shaped bone plugs. Before heading home, the hunter might adjust his snow goggles, made from bone or ivory. They fit like a ski mask and had extremely narrow horizontal slits to minimally admit the blindingly bright light.

Museums house many such artifacts, some common, others rare—bone body armor, for instance. Scores of 2-by-8-inch plaques of thong-bound bone, likely cannon bone, formed vests to provide protection from chin to midthigh.

In more temperate climates, indigenous people for thousands of years have made turkey callers out of turkey wing bones. After a quick Internet search, you can see videos of them at work. I bought a set of dry, clean turkey wing bones online and made my own caller. Each of the three tubular bones is slightly tapered. By cutting the ends off of all three and cleaning out the sparse interior spongy bone, I was able to assemble the trio into a small fanfare trumpet, except that one calls turkeys with pursed lips and smooching inhalations. The resulting yelps resound through the forest—well, in my case, the house. If you are just minimally interested in crafting a killing accessory from bone, you might try a chicken wing bone caller instead. I did. It works. I mean it squawks, but no chickens appeared.

As we now know, people repurposed bones both out of necessity and out of availability. Take whaling as a case in point. It was a huge industry in the nineteenth century when whale oil was in demand for lighting. In the process of harvesting the blubber, the seamen cast aside unfathomable tons of bone but saved some for carving. With time on their hands between hunts, ingenious whalers repurposed this readily available crafting material into personal items and tools. Although some of these objects could have been made out of metal at that time in history, whalebone was at hand, did not rust, and required only a pocketknife for shaping. Practical tools so produced included sail needles and seam rubbers, which were used by sailmakers to rub the canvas flat once sewn, and also belaying pins and fids. For landlubbers, the first were stout rods around which rigging lines (ropes) were secured. In modern times, think cleats. Fids were stout, conically shaped awls used to hold open knots and holes in canvas and to pry loose strands of line when splicing. Though fictional, it is worth noting a literary reference to bone as a hunting accessory: Moby Dick chomped off Captain Ahab's leg, and the ship carpenter fashioned a replacement out of, quite fittingly, whalebone.

For all the purposes that humans have found for bone, only one that originated at civilization's outset has endured to the present—bones in the diet. Maybe from watching wild beasts crunch them open and feast on the soft interior, early humans learned to access this delectable core by splintering a large animal's leg bone between rocks. Archaeologists arrived at this conclusion based on discovery of shattered and charred animal bones in firepits and rubble heaps at numerous ancient dwelling sites.

Today, to ease access to the marrow, butchers can saw bovine leg bones either into short cylindrical segments or into longitudinally divided halves. Stacked on end in a baking dish, the cylinders resemble little tree stumps, and the halves look like miniature canoes. After about 20 minutes in a hot oven, the marrow becomes soft and creamy and spreads easily on toast. Gourmands love its rich, buttery texture and taste.

To flavor soups and sauces, many cooks traditionally use broth derived from simmered bones of fish, fowl, or four-footed critters. In recent years, bone broth bars have appeared and offer patrons a non-caffeinated, nutritious alternative to coffee or tea. In the more health conscious of these establishments, the proprietors tout these elixirs for their filling, cleansing, and detoxing capabilities. Bone broth is also the central ingredient for pho, Vietnamese noodle soup.

But marrow and broth are merely bone's interior content; what about eating bone itself? Rodents gnaw on dry ones, presumably to fulfill their calcium and phosphorus cravings. Canned salmon and sardines both include some small bone bits—a little crunchy but entirely palatable.

In several Middle Eastern and Asian cultures, the same holds true for eating a whole small bird, whose carcass is about the size of a human thumb. I experienced this years ago while on a hand surgery cultural exchange in China. My banquet hosts advised me to chew slowly to crush the bones of a whole deep-fried fledgling sparrow to avoid impaling my palate with fracture fragments. Since I kept that warning foremost, I do not recall how it tasted—just crispy.

In France, they have taken eating an intact, bite-sized bird—the ortolan bunting—to an extreme. Traditionally, gourmands cover their heads with their napkins during the experience for some combination of the following reasons: hiding shamefully from God, retaining the aromas of the delicacy, and shielding tablemates from the act of beak spitting. In his book *Medium Raw*, Anthony Bourdain writes, "With every bite, as the thin bones and layers of fat, meat, skin, and organs compact in on themselves, there are sublime dribbles of varied and wondrous ancient flavors: figs, Armagnac, dark flesh slightly infused with the salty taste of my own blood as my mouth is pricked by the sharp bones."

This practice is now banned in France because it was driving the bunting toward extinction. A similarly disturbing and age-old tradition in the Far East is making and ingesting tiger bone wine. The bone is added to rice wine and aged. Purportedly, the product has widely curative powers, but much to the detriment of the wild tiger population.

Less startling, far more common, and spanning back to the beginning of civilization is man's (more likely woman's) uses of bone for gathering, preparing, serving, and storing food. For instance, various widely separated cultures used adze-like root picks for in-soil

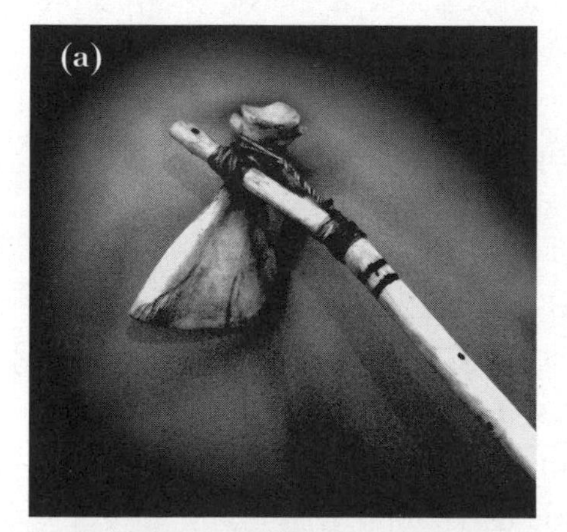


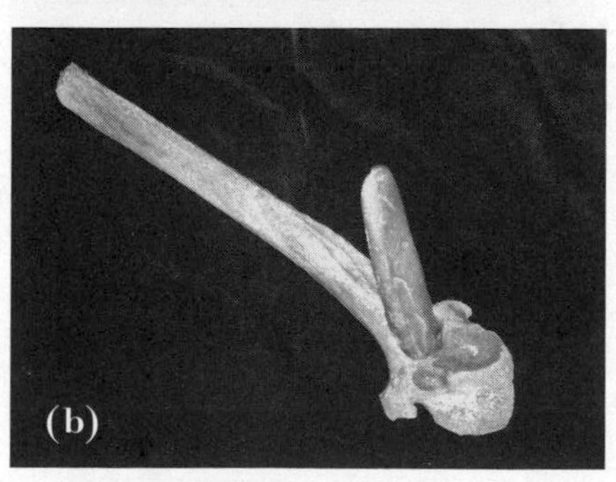


*Native Americans and many other indigenous cultures repurposed bones for agricultural purposes. These include a hoe made from a bison shoulder blade (a) and a root pick consisting of a stone blade and a bison vertebra handle (b).* (A) UNIVERSITY OF NEBRASKA STATE MUSEUM; (B) KING ROSS.

foraging. Some picks have a curved bone blade, likely a rib, and a wooden handle. Others have a bone handle and a stone blade. For harvesting, deer jawbones became sickles. Later, beasts of burden pulled plows, which were tipped with spikes of long bones to tear the soil open.

People shaped the large flat shoulder blades of elk and bison into hoe blades and also sharpened them along one edge for slicing soft plant materials, so-called squash knives. Also, serrations on a shoulder blade's long edge converted it into a steak knife.

Bone tools also aided preparation of other foods. Slate and flint knife blades fixed to bone handles were common. For people living near the water, flattened and sharpened segments of bone became eel splitters and fish skinners. Skulls served as mortars. Even when the mortar was stone, the pestle could be bone. Some Native Americans did not bother deboning small birds, but rather pounded the whole bird with a mortar and pestle, cooked the mash, and spat out the bone bits during dinner, perhaps anticipating Anthony Bourdain's gastronomic ecstasy.

Baskets and round boxes crafted from strips of softened bone provided storage for foodstuff, as did saltshakers and nutmeg graters on a smaller yet grander scale. For example, beginning in the mid-seventeenth century and extending into the nineteenth, the gentry carried pocket-size and ornately crafted nutmeg graters. Silver, ivory, or bone in composition, the graters provided fresh spice to flavor alcoholic punches and ciders and to protect against the plague. Nutmeg graters would have been right at home with a fine setting of bone china.

For early people living closer to the earth a bowl of broth might have come in a hollowed-out dolphin or whale vertebra, a turtle shell, or maybe a skull. Eating broth with a spoon gradually overtook slurping. Forks followed. Initially, both were made entirely of bone, later just the handles. In other parts of the world, chopsticks made of bone served similar purposes. Specialized utensils made from bone included apple corers, made from stout cannon bones, and long slender marrow spoons, designed to garner every last morsel of delectable from a bone's interior—a case of using bone to eat bone to nourish bone.

Another basic need of primitive humans was shelter. In regions where caves, rocky overhangs, and construction-sized timbers were scarce, one solution was to build a bone home. From over 70 sites in Ukraine and Siberia, archaeologists have found remains of dwellings up to 26 feet in diameter made of stacked mammoth bones, now collapsed. Likely they were lashed together and then covered with animal hides to turn the wind.

The strangest bone home is likely what *Ripley's Believe It or Not* calls "The World's Oldest Cabin." The structure itself stems from 1932, but it consists of dinosaur fossils, hence the "old" claim. An amateur paleontologist began collecting material from a bone bed in Wyoming several decades after Cope and Marsh had warred over prehistoric treasures exposed on the same bluff. The dabbler planned to assemble a complete dinosaur skeleton as a roadside attraction for travelers needing gas. When that proved impossible, he took his cache of nearly 8,000 assorted fossils and built a cabin, which stands today, near Medicine Bow in southeastern Wyoming.

Better documentation is available for bone homes used by Inuit tribes in northern Canada. They used whale jaws and ribs to frame dome-shaped, semi-subterranean domiciles. The jawbones were slender throughout their 18-to-23-foot length (see p. 37) and perfectly curved to arch overhead. Around the edge of the pit, 16 or more of these were anchored at their bases by large stones and then lashed together centrally to form a circular dwellings up to 15 feet across. The builders sequentially covered this skeleton with hides, moss or turf, and then snow to provide a snug interior. Capping it off, a cylindrical, hollowed-out whale vertebral body became the chimney flue.

The Inuits also used bone as construction accessories. Large penis bones became tent stakes. Machete-like snow knives, originally crafted from bones, were used to carve blocks of snow for igloo construction. For added protection once the snow blocks were in place, women and children used shovels crafted from caribou shoulder blades to heap snow over the dome.

Smaller animals in far milder climes have contributed ribs to form 6-inch-long thatching needles. The roofer sutured thatch to the roof beams with cord using these scary, clawlike tools—sharply pointed on one end and containing an eye on the other. Much smaller bone needles were used in many early cultures for sewing clothing; and although the first stitched garments are long gone, archaeologists estimated when the craft began after finding 20,000-year-old bone needles.

Soon after primitive humans were satiated and sheltered, they may have stared at one another and noted how crude their loincloths and ill-fitting capes looked, both made from hides. This was the start of the fashion industry, and bone played a central role from the beginning.

Beamers were cannon bones with one surface of the shaft removed. This presented two long sharp edges useful for defleshing and scraping hides. Shoulder blades with small serrations along the thin edges served the same purpose. A shoulder blade with a 1-to-2-inch central hole became a thong stropper, which allowed the user to draw a strip of skin back and forth through the opening to soften the hide and remove any residual flesh. Early seamstresses used bone awls, which were long, polished spikes that made sizable holes through which the seamstress passed a thong on a blunt bone needle to create the latest fashions.

The origins of textiles are lost in prehistory. Discovery of both thin needles suitable for poking minuscule holes in fabric and twisted threads of flax implies that fabrics—woven, netted, or knitted—are at least 10,000 years old, maybe much older. Over the course of time, bone has been central to many aspects of fabric manufacture, the beginning of which required spinning some plant- or animal-derived fiber into thread or yarn. This often entailed use of a distaff and spindle whorl, both made of bone. It was typically women's work, and many Renaissance artists have romanticized a

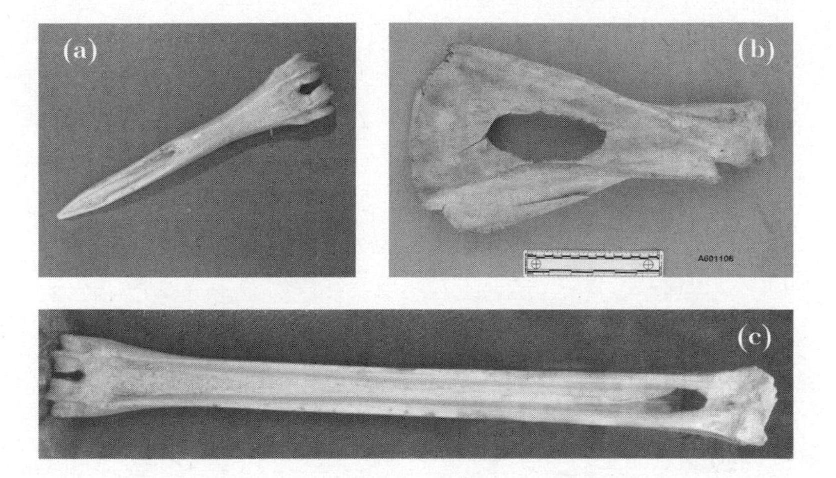


*Bone tools were integral for hide preparation. (a) This cannon bone awl penetrated the leather prior to stitching it with a thong. (b) The hole in this shoulder blade served as a thong stropper. A strip of hide passed back and forth through the hole became supple and flesh free. (c) This cannon bone beamer scraped flesh from hides and softened them.* (A) MUSEU DE PORTIMÃO; (B) NATIONAL ANTHROPOLOGICAL ARCHIVES, SMITHSONIAN INSTITUTION [ITEM A601108]; (C) MUSEUM OF NATIVE AMERICAN HISTORY.

lovely lady with a long rod (distaff) supporting a mess of flax or wool fibers above her shoulder and a dangling rod with a rounded weight (whorl) spinning at her knee. The whorl's momentum spun a thread as the woman paid out fibers from the distaff and twisted them between her fingers.

Weaving also benefited from the availability of bone tools. For looms, this use included shuttles, of course, but also weaving swords, which the Vikings made from long thin slats of whale jawbone and used to pack the weft tightly after the shuttle had carried the thread through the warp.

Card weaving, also known as tablet weaving, is a loomless form of fabric manufacture. It originally entailed use of small, thin squares or triangles of bone, bark, or horn through which selected threads of the warp were passed. Then by rotating one or more of the stacked

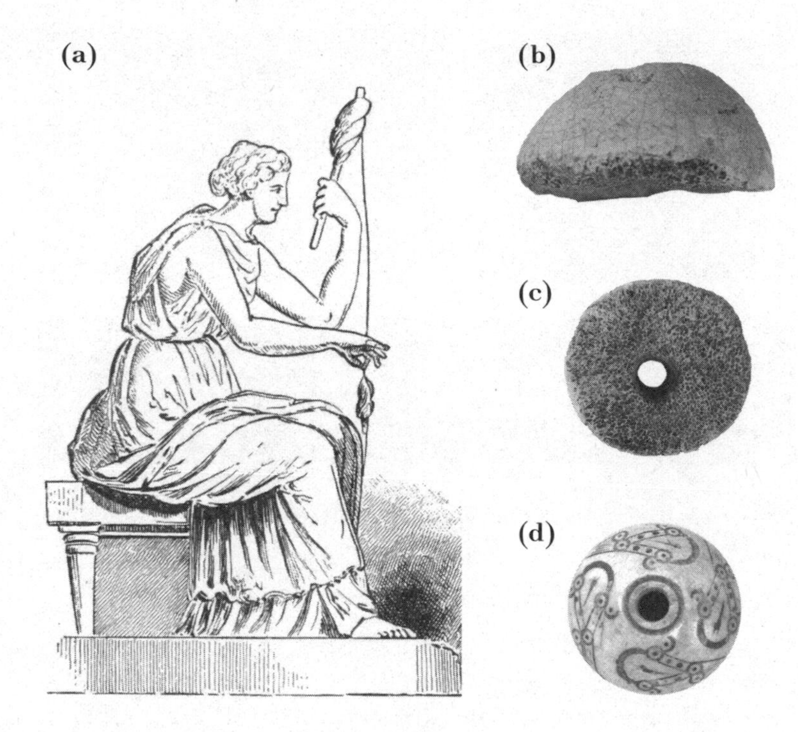


*For 3,000 years prior to the development of the spinning wheel, converting plant or animal fibers into thread and yarn began by placing the disorganized jumble of fibers on a rod, often made of bone, held in the left hand. The right hand pulled out fibers a few at a time and twisted them (a). A weighted whorl maintained the spinning thread's momentum and was often made from the "ball" of a cow's hip joint. It was perfectly sized and shaped for the task and was left plain (b, c) or inscribed (d).* (A) S. N. DEXTER NORTH, "THE DEVELOPMENT OF AMERICAN INDUSTRIES SINCE COLUMBUS. V. THE MANUFACTURE OF WOOL." *POPULAR SCIENCE MONTHLY*, 39 (MAY–OCTOBER 1891): 176–95; (B, C) PORTABLE ANTIQUITIES SCHEME; (D) METROPOLITAN MUSEUM OF ART.

cards, different warp threads were lifted for each transverse pass of the weft. Card weaving lent itself particularly well to the formation of narrow straps and belts. Various archaeological sites have yielded bone cards and an occasional fabric band so produced. Cards made of wood or horn were less likely to withstand the ravages of time.

Although tablet weaving remains a popular craft today, the cards are more likely plastic or homemade from playing cards.

Other bone instruments for intertwining thread include netting gauges, knitting needles, crochet hooks, lucets, and bodkins. A lucet is a two-pronged, handheld tool, sort of a miniature loom useful for braiding drawstrings and shoestrings. Ones made from bone are often found among Viking artifacts. Lace makers wound their threads on bodkins (elongated spools), which they used in matched pairs. To reduce confusion regarding which thread next went where, each pair of bodkins had identically carved heads. If one spool became misplaced, the other was an odd bodkin.

Stitching fabric panels together became more versatile when metal needles replaced those made from bone. Nonetheless, bone remained integral to the process. Cylindrical bone cases stored the needles, and bone thimbles pushed the needles through heavy fabrics. If a seam needed dismantling, a raccoon penis bone, naturally curved and then ground to a sharp point on one end, served well.

When it came time to fasten clothes and fix hair, bone proved to be durable and universally available. Bone pins became progressively stylish, yet they ultimately yielded their position to buttons

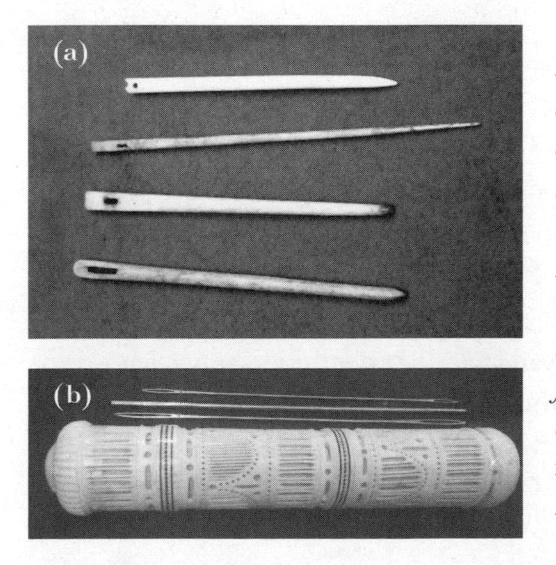


*Bone tools aided the development of sewing, crocheting, knitting, and net making. (a) The sewing needles are from the Roman era in the Bordeaux region. (b) The box for netting needles was made from bone in nineteenth-century England.* (A) Musée d'Aquitaine; (B) Victoria and Albert Museum.

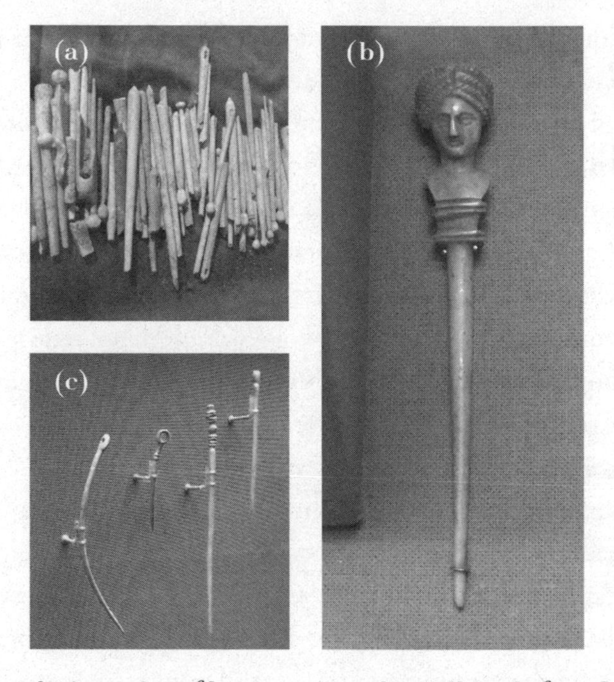


*Prior to the invention of buttons, pins—here all made from bone—fixed hair and also secured loosely fitting hides, and later woven fabrics, on the wearer's body: (a) from Roman occupation of Spain, about 200 BC to AD 500; (b) from Roman London about AD 200; (c) from Japan, Jomon era, 2000 to 400 BC.* (A) PALACE OF THE COUNTESS OF LEBRIJA; (B) BRITISH MUSEUM; (C) TOKYO NATIONAL MUSEUM.

and barrettes. Bone belt buckles and strap ends also were both functional and fashionable. Netsukes became a high-art form for Japanese button-like toggles. This cord stopper, which secured a pouch containing the owner's personal belongings to the sash on his robe, was at first strictly utilitarian, but over time it became highly refined and reflected Japanese life and folktales. Many netsuke collections have specimens made of bone, although most are ivory.

Bone also helped clean and maintain clothing. The whalers carved clothes pins from bone, and soldiers kept button polish off their uniforms by using highly engraved, bone-crafted "button sticks," which were slats of bone with a central slit connected to a button-sized hole at one end. The well-appointed military man placed a button through the hole, slid the stick under the brass fastener to isolate it from the garment, and polished away.

Not only did early humans use bones to feed, shelter, and clothe themselves, they also used them for purposes of health, comfort, and convenience. Museum collections include syringes made entirely of bone, ornately decorated fine-tooth combs, complete sets of dentures, and eyeglass rims, all certainly luxuries in their time.

More mundane are the aids that were used to facilitate humdrum daily living. These objects reflect each culture, its particular needs, and the versatility of bone in addressing those needs. I have found a panoply of utilitarian, often artfully crafted, and domestically useful bone items in museum collections and at auction house websites. These include (take a deep breath if you are reading out loud) candlesticks, dishmop handles, glove stretchers, cuticle pushers, combs, razor handles, money purses, face wipers, drinking tubes, tweezers, shoehorns, canes, umbrella handles, boot jacks, boot sole creasers, corkscrew handles, chairs, eye cleaners, toothbrushes, ear scoops, scratching sticks, and louse killers. Even though the candles, gloves, bottle corks, and lice to which users applied these tools are long gone, the bone implements remain to document humankind's cultural evolution. At times, bone's applications have gone far beyond strictly utilitarian applications.

*Chapter 15*

# BEGUILING BONES

At this point you may be sufficiently enraptured by bone's use in everyday life that you want to buy some finely crafted items for your intense study and personal enjoyment. But before doing so, make sure that you can distinguish bone from ivory in case you should come across an irresistibly beautiful crafted white object in an antique store. Can you trust the dealer's word regarding its composition? Museum curators and the US Fish and Wildlife Service also have vested interests in differentiating bone from ivory. How can you tell them apart?

Even from afar, consider the object's size and shape. If it is curved, 3 feet long, and tusk shaped, it's ivory. If it is short, stout and conical, and scrimshawed by your nineteenth-century sailor-ancestor, it's a whale's tooth. If it is at least 3 inches square and flat, it came from a whale's jawbone. A sperm whale's jawbone can be 25 feet long; and at the end where it contacts the skull, the bone is thin and wide and can yield a panel as large as 7 by 12 inches. No tusk can match these dimensions.

Then, disregarding size and shape, consider its context. If the object came from an eighteenth-century royal treasury and is encrusted with emeralds, it's probably ivory. If it is flat and fully painted with a scene or portrait, it's more likely bone, since painting over precious ivory would be akin to wrapping engraved gold trays in aluminum foil. If the treasure was unearthed from a

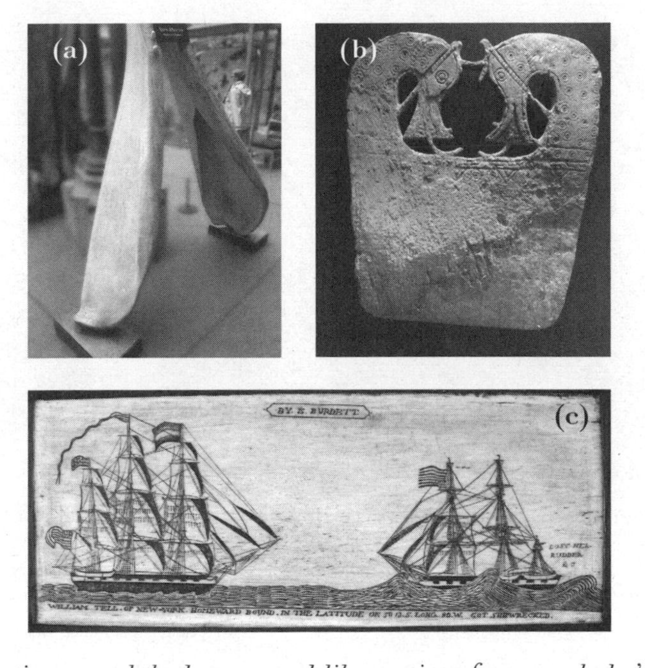


*(a) Artisans used the large, panel-like portion of sperm whales' jawbones for painting, etching, and carving. (b) For instance, this ninth century AD Viking plaque was perhaps used as a food tray or chopping board. It measures 8 by 9 inches. (c) In 1831, Edward Burdett scrimshawed sailing vessels onto this 6½-by-12½ inch section of jawbone.* (A) OXFORD UNIVERSITY MUSEUM OF NATURAL HISTORY; (B) BRITISH MUSEUM; (C) COURTESY OF THE NEW BEDFORD WHALING MUSEUM.

Native American mound in Illinois, far from elephant and marine mammal habitats, most likely it is bone, although it might be mammoth tusk.

Seen best with a magnifying lens, ivory has a pattern of cross-hatched lines, known as Schreger's lines, and their angle of intersection differentiates elephant from mammoth ivory. Bone does not have such lines; rather, its surface will have tiny dark dots and parallel dashes. In life, these were the minuscule passageways, Haversian canals (seen on p. 32), through which blood flowed to nourish the bone cells.

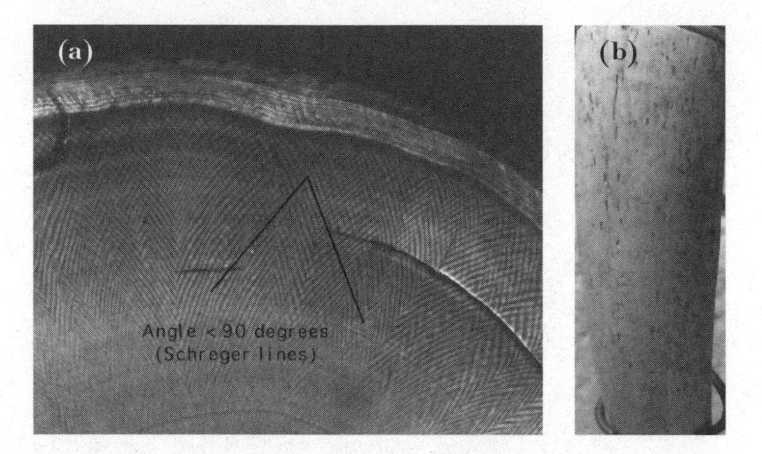


*(a) Under 10× magnification, ivory shows a distinctive pattern of crosshatched lines. (b) This is a close-up of the vomit stick depicted on p. 177. It shows that bone has pits and streaks usually visible without magnification. These bone markings are canals containing blood vessels and are seen end-on on p. 32.* (A) US FISH AND WILDLIFE SERVICE; (B) MUSEUM OF NATIVE AMERICAN HISTORY.

Finally, there is the hot pin test. A pinpoint heated red-hot and touched immediately to an inconspicuous area on an ivory object will not affect it. On bone, the hot pin is said to produce an aroma of "burning hair."

Perhaps it is better to have a professional, one who does not have a financial, legal, or emotional stake in the answer, appraise the treasure. In the past, I can imagine that many bone objects were promoted as ivory to unsuspecting buyers. With few and diminishing exceptions, there are stringent bans on the international sale of ivory. In a growing number of states, including California, New Jersey, New York, Washington, and Hawaii, trade in ivory is banned altogether. So now dealers might try to pass off their no-longer-marketable ivory objects as bone.

I wondered if that was the case when I saw a graceful back scratcher in an antique shop just a bone's throw from the Louvre. The dealer said that it was old and made of bone and named a hefty price. It was finely crafted, quite smooth to the touch, and without any dark dots or streaks. Based on its size and shape, it could have been either bone or ivory. My uncertainty combined with a mental image of the US Fish and Wildlife Service marching me into the slammer led me to leave it alone. "I'll have to find another gift. Do you have anything else in bone? You sure it's bone?"

I am absolutely sure that the "knucklebone" gaming pieces I recently dug up in my backyard are bone because six months earlier I buried the goat legs there that I had obtained from a butcher. In the interim, soil organisms had cleaned the bones completely, leaving me this skeletal sequence from knee out: leg bones, the prized knucklebone (one of the hock or anklebones), a cannon bone, and an assortment of toe bones. The goat's knucklebone is about the size of a hard candy and is roughly rectangular. Four or five fit easily in one's palm, and it is almost impossible not to jiggle them and roll them out—a thrill of touch, sight, sound, and anticipation. For thousands of years, probably beginning in ancient Egypt or in what is now western Turkey, others have enjoyed the same simple pleasure.

Each of the knucklebone's four long sides is uniquely shaped, so the odds of it landing on any given side are different. (The ends are too rounded for it to stand up straight.) Without concern for how they landed, children played jacks with them or tossed them up and tried to catch them on the backs of their hands. By attributing different values for each landing surface, adults used them for gambling, from whence comes the expression, "Roll the bones." Fortune-tellers attributed meaning to the landing positions both of the individual bones and in relationship to one another.

The popularity of playing with knucklebones is evident from the frequency with which they appear in the archaeological record, both as bones and as identically shaped gaming pieces crafted from other materials. Through the ages, sculptors and painters have captured knucklebones in action.

Cubic dice, many originally carved from bone, standardized the odds of a gaming piece landing on a given side. Other shapes

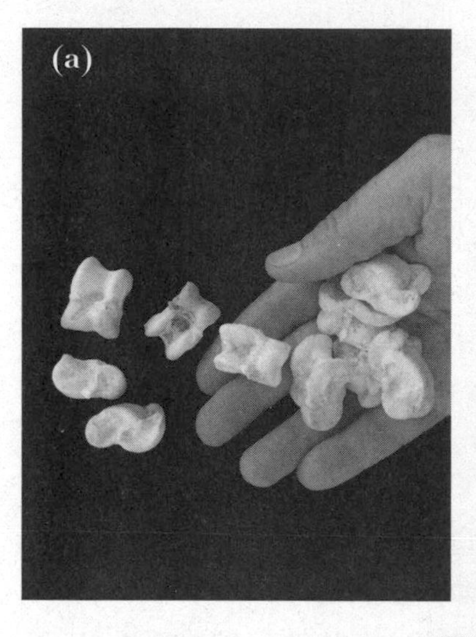

*(a) Anklebones from goats and sheep, known as knucklebones, have been widely used since ancient times for gambling, fortune-telling, and games. (b) The woman in this 1734 painting is playing "jacks" with a ball and four knucklebones on the table. (c) The subject of this Greek vase from about 350 BC is games of love. The point was to toss the knucklebones up and catch them on the back of the hand. The nymph is gambling with a satyr.* (B) BALTIMORE MUSEUM OF ART; (C) WALTERS ART MUSEUM.

ensued, and I might have found solid geometry more interesting had I known then that some dice were tetrahedrons (4 sided), icosahedrons (20 sided), and nearly every sidedness in between. For those seeking an alternative to rolling the bones, spinning a die-type top, variously named a teetotum or dreidel, was also common and may have originated in ancient China or Japan or both. Bone has also been the material for early versions of Pick-Up Sticks (also known as jackstraws and spillikin) and for domino and mahjongg tiles.

Casting the die allowed gamers to compete for cumulated scores, but more visually interesting were board games. Backgammon and checkers both originated at least 5,000 years ago in the Middle East, chess got its start almost 2,000 years ago in India, while cribbage, from Britain, is only about 400 years old. Ivory and bone playing pieces for these games turn up frequently in archaeological sites but prove difficult to date because of their similarity over time.

By AD 1400, the Embriachi workshop in Italy, widely known for producing elaborately carved bone altarpieces and jewel boxes, was also turning out double-sided game boards. The bone inlay was patterned for backgammon on one side and chess on the other.

Anglo-Saxon gamers carved chunky chess pieces from whalebone. In medieval times, playing chess was considered not only fun but also gentlemanly and chivalrous. It pitted the competitors' strategic and tactical abilities within a framework of rules and sportsmanship along with the art of winning or losing gracefully. In this context, ornately carved chess sets appeared, made from a wide variety of materials, including bone. Although the native shape of bone lent itself easily to cubic and disk-shaped gaming pieces, craftpeople had to apply some ingenuity to form bone into elaborately shaped chess pieces. They made flat bases, thin stems and finials, and thick, hollow midsections from separate sections of bone and then threaded them together.

Native Americans devised unique bone-centric games and crafted their own playing pieces, from a mountain lion when possible. The animal was respected for its cunning, a trait that the crafter hoped would transfer to his game.

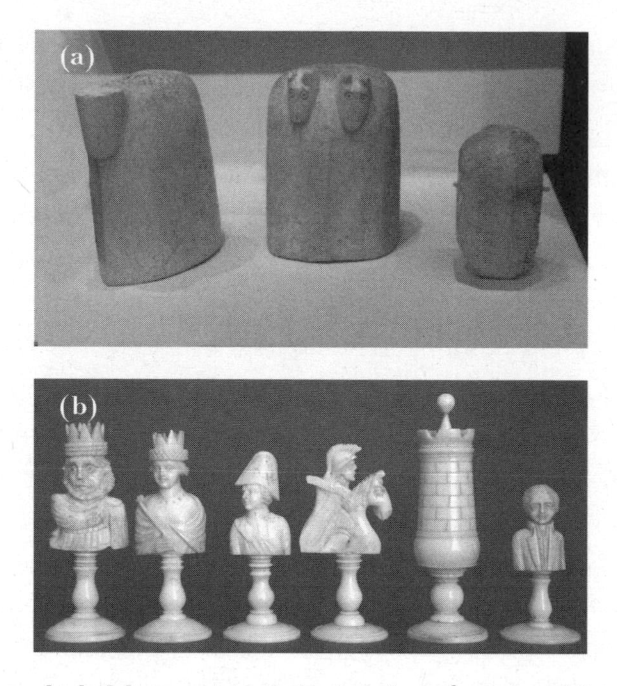


*(a) Carved whalebone provided playing pieces for a chess-like game from about AD 900 in Anglo-Saxon England. (b) This far more refined bone chess set was carved in Dieppe, France, about 800 years later.* (A) BRITISH MUSEUM; (B) THE JON CRUMILLER COLLECTION.

Pastimes for First Nations gamers also included bone-based forms of the cup and ball game and ring and pin game known the world over. They all require hand-eye coordination to catch a tethered ball in a cup or to flip a ring onto a spindle. In wood-sparse Greenland, a rabbit's skull with extra holes drilled into it substituted for the ball. In another Inuit variation, hollowed-out deer toe bones strung together on a leather thong were the objects to capture.

Dolls carved from bone were common in ancient times. The simplest were "stick dolls"—slender, no arms. More complicated dolls had arms and legs joined to the torso with pins. All were small, reflecting their osseous origins. Museum collections label some "figurines," and since none came boxed with descriptive labels, it is now up to imagination to decide which ones were toys, which ones were objects of beauty, and which ones had spiritual powers.

While Inuit adults used snow knives made of bone for igloo construction, their children used smaller versions, called story knives, to illustrate their fantasies in the snow, scrape them away, and start over. No need for Etch-A-Sketch.

Children from around the world played with other bone toys, necessarily small because of the material's constraints, but widely diverse according to the maker's ingenuity. Toys that have survived hard use and that now reside in museums include teethers, doll furniture, whirligigs, Scrabble-like letter tiles, and miniature wagons, sleds, and kayaks.

Bone-facilitated amusements were not exclusively sedentary, low-energy activities. Medieval English and Swedish archaeological sites have revealed ice skates made from longitudinally flattened cannon bones. William FitzStephen, who chronicled twelfth-century London life, described the skaters: "Some of them fit shinbones of cattle on their feet, tying them round their ankles. They take a stick with an iron spike in their hands and strike it regularly on the ice, and are carried along as fast as a flying bird or a bolt from a catapult."

Even before the advent of simple games, primitive folk chanted, sang, and played instruments, the first of which may have been a couple of charred mastodon ribs. Some cave dweller clacked them together and smiled. Instrumental music was born.

Playing the bones has continued through time. Shakespeare knew of the art. In *A Midsummer Night's Dream*, Bottom commands, "I

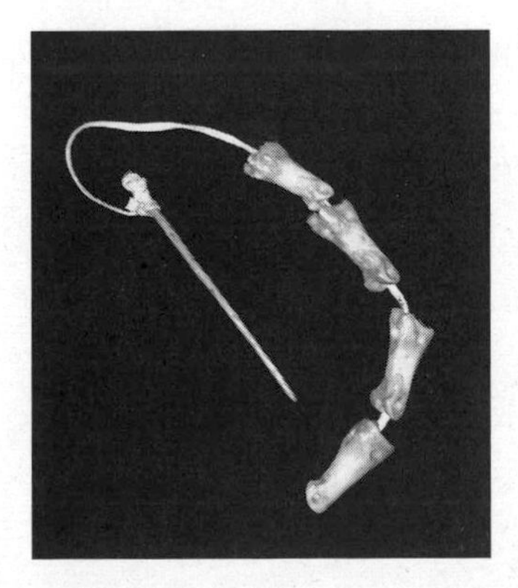

*Arapaho Indians on the Great Plains played their version of the ring and pin game using a bone needle and four deer toe bones.* NATURAL HISTORY MUSEUM OF LOS ANGELES COUNTY.

*The origins of rhythmically clacking rib bones together are lost in time, but the art is memorialized in this 1856 painting by William Sidney Mount.* MUSEUM OF FINE ARTS, BOSTON. BEQUEST OF MARTHA C. KAROLIK FOR THE M. AND M. KAROLIK COLLECTION OF AMERICAN PAINTINGS, 1815–1865.

have a reasonable good ear in music; let us have the tongs and the bones." In the visual arts, William Sidney Mount, best known for his depictions of everyday life, painted *The Bone Player* in 1856.

Folk musicians still play the bones, and there are commercially available sets made from wood. Should you want to play the bare bones yourself, enjoy some barbecued spare ribs, clean them up, and clack away as your ancestors did perhaps 40,000 years ago.

Another extremely ancient percussion instrument consisted of a bone notched transversely along its length. When stroked with another bone, the rattling tempo could be faster than obtainable by clacking ribs together.

Another musical instrument/noisemaker/toy/communicating device with ancient roots is the bull-roarer, a flat plate of bone attached to a long thong or cord. When swung overhead, it makes a low-pitched vibration that varies according to the speed and plane of the swing and the length of cord. These appear in archaeological sites as old as 18,000 years. Turning up with greater frequency are small animals' cannon bones with central holes drilled through. The cords are absent, but these were likely "buzz bones," activated to spin and hum by pulling on a twisted cord that passed through the hole.

To see the myriad ways that more advanced cultures have used bone to make music, visit a musical instrument museum, which I did recently in Phoenix. It appeared that every country in the world and nearly every ethnic group was represented. Sure, wood and metal constitute the principal materials for many of the instruments, but boneheads will not be disappointed. Smallish but integral osseous components include violin nuts, saddles, bridges, and bow-tip plates; bagpipe fittings; tuning pegs for various instruments; and guitar nuts, picks, and slides. For other instruments, bone inlay ornamentation contrasts nicely with wood. Armadillo armor is bone, and several enterprising instrument makers turned the creature's hard "shell" into the bodies of ukulele-type instruments. My favorites were the ones where bone constituted the entire instrument. For the percussion section, the slightly loose teeth in a horse's

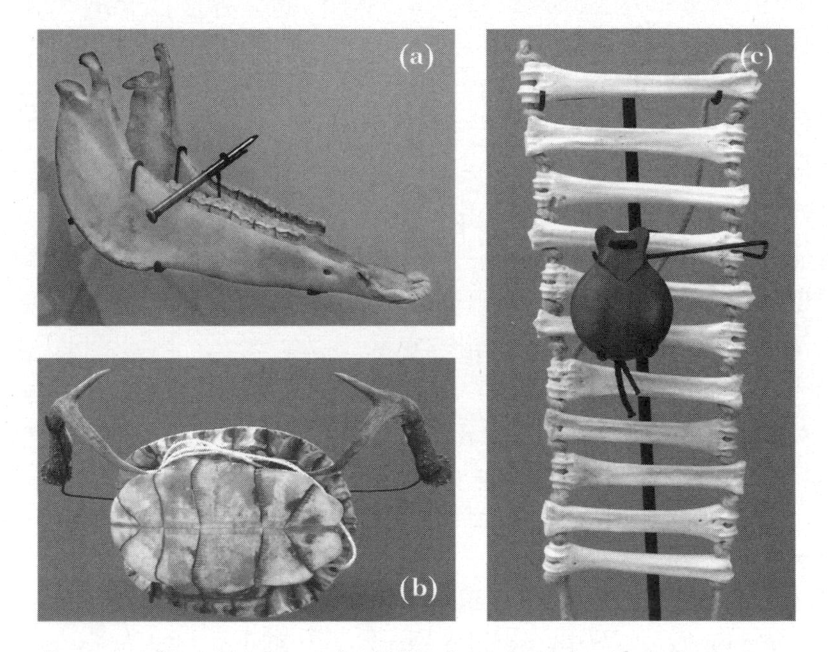


*Because of its ubiquity and solidity, bone lends itself nicely to the fabrication of musical instruments. (a) The teeth in the horse jawbone are loose enough to rattle, and different effects are achieved either by stroking the teeth with a spike or by holding the jaw on its narrow end and tapping the wide end as if playing a tambourine. (b) The musician taps this Mexican turtle shell with deer antlers. (c) A Spanish "bone scraper" is made of cannon bones and is rattled with castanets.* MUSICAL INSTRUMENT MUSEUM.

jawbone rattle when slapped. (I tried a replica in the gift shop.) Also a breastplate made of goat cannon bones begs to be strummed with castanets, just like a washboard.

Not to overlook woodwinds, in great abundance were bone flutes from around the world. They demonstrate a wide spectrum of complexity, size, artistic embellishment, and antiquity. Some "flutes" are as old as 35,000 years, yet the experts argue sometimes whether a bone fragment with several holes in it was human-made or the result of a strong carnivore's bite. More recent hollow bones are clearly musical instruments. Some are end-blown, like modern-day recorders. Others are side-blown, like piccolos. A few have multiple tubes. Certainly, hollow bones have well served humankind's interest in entertainment and spiritual fulfillment.

If your search for osseous musical instruments does not yield a musical instrument museum, try a whaling museum. The whalers loved music as much as they liked crafting bone. The New Bedford Whaling Museum exhibits a banjo made almost entirely from bone and a violin with a bone fingerboard and tail, and the Nantucket Historical Association displays a bone music stand of similar vintage.

If music alone insufficiently soothed the savage beast, people could turn to tobacco, snuff, and opium, aided by bone accessories, of course. Pipes, tamps, matchboxes, snuff boxes, and opium spoons, often ornately carved, attest to humans' timeless attraction to these habits, which encompassed not only social exchanges but also religious practices.

In addition to modifying bone for spiritual, practical, and diversionary purposes, humans have also fashioned bone solely for aesthetic delight. Its solid, cool, pale surface is appealing to view and touch. Its moderate degree of hardness, between that of wood and stone, makes it relatively easy to craft yet leaves it durable for centuries. Extolling bone's aesthetic properties may lead one to ask, "How do bones get from their living state—muscle covered, cartilage capped, and fat filled—to the inert, dry, pleasing state in which they are widely exhibited and valued?"

When a carcass remains exposed to the elements, it takes a year or two for the soft tissues covering the bones to weather away and for microorganisms and small insects to feast on and completely remove the fat from the bones' interiors. (The tiny scavengers get in and out through the same small channels that the blood vessels used during life.) By happenstance, craftspeople may come across such naturally cleaned and dried bones on the beach, desert, or forest floor, in which case repurposing can begin immediately. During this slow transformation, however, scavenging animals may carry the bones off or destroy them.

How do craftspeople and museum staffs hasten the transition and ensure that the bones will be there for modification or display when the cleaning is complete? The three approaches are the same ones used by the US Marines: land, sea, and air. Each path has its advantages and disadvantages. For bone preparation, a preliminary removal of the skin, muscles, and innards accelerates the process, but patience will even replace this grizzly step should one desire.

Land: burying the remains (deeply enough to prevent scavenging) has the advantage of being stink free. Before interment, I wrap my treasures in nylon mesh so I can easily retrieve even the small bones in due time. This process usually takes 6 to 12 months. Presently I am processing a squirrel (road kill), a moose cannon bone (hunter friend in Alaska), a chicken (Whole Foods), and a small pig (Hawaiian luau).

Sea: Placing the remains in a bucket of water, or in a huge vat if you are preparing an elephant, greatly hastens the transformation from messy to beautiful. This method is extremely offensive to the nose, however, and so should be done far from habitation. It is best to change the water every few weeks, ideally during a windstorm or when you have complete nasal congestion. (The whalers avoided this problem by towing the to-be-worked jawbone behind their ship for a month or two.)

Air: This method is not quite so stinky but still needs to be done outside and certainly not under an open window. The bones should

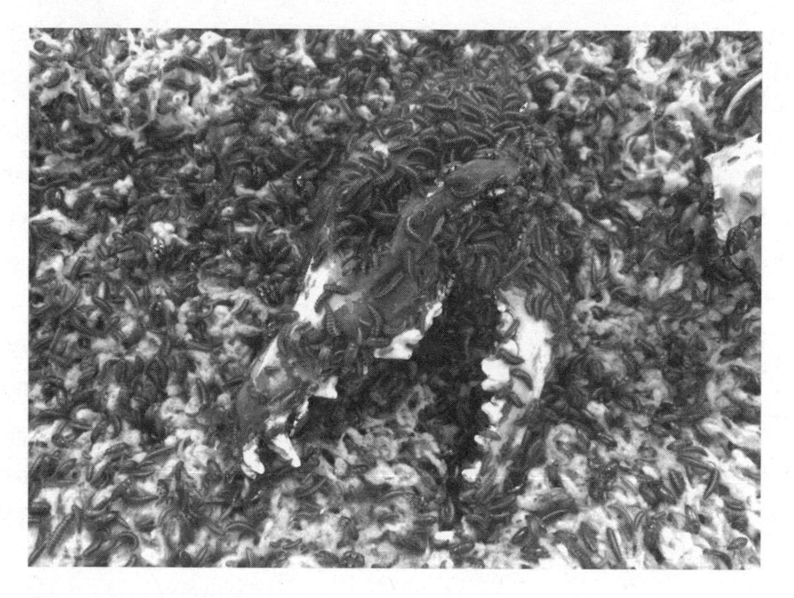

*Dermestid beetles thrive on decaying flesh but leave bone entirely alone, thus endearing themselves to preparers of skeletal displays, here the skull and jawbone of a fox.* MUSEUM OF OSTEOLOGY.

be wrapped in window screen or hardware cloth to prevent predation. (I and others keep such weighted-down preparations on our housetops, typically unbeknownst to our spouses.)

To accelerate air transformation, university and museum laboratories and commercial preparers often employ dermestid beetles. They love carrion and quickly nibble bones clean, recruiting their smallest larvae for the inside job. Should you wish, the beetles are available for both viewing and purchase on the Internet. Before considering ownership, understand that they need regular feeding, so they are not practical for a one-time harvesting of a Thanksgiving turkey skeleton.

After the cleaning by whatever means is complete, a hydrogen peroxide bath will bleach the bones pearly white. There are, however, two ways to absolutely ruin hard-earned specimens. Household bleach permanently softens and pits bone, making it unpleasant to handle and unsuitable for crafting or display. Boiling also renders bones useless for crafting. Heat drives the marrow fat into the dense, normally fat-free outer portions. The relocated fat is inaccessible to wee beasties of all sorts and leaves the bones permanently greasy.

Early humans, of course, did not dry and bleach bones for secondary use. They just picked them up ready to go and modified them to meet survival needs for shelter, clothing, and food preparation. Eventually, there was extra time, energy, and dry bones left for crafting, and humans began repurposing them to interpret their world for aesthetic and spiritual reasons. Modern discovery of prehistorically crafted bone, therefore, provides a reflection on the human soul and helps establish a developmental time line for the origin and development of abstract thought and spirituality. In this context, anthropologists have a heyday interpreting the meaning or lack of meaning of any given discovery and may reframe their concepts of human mental development and cultural evolution as new findings surface. They can trace general trends by piecing together bits of information in the archaeological record gleaned from sites worldwide and from over at least 40,000 years, which was the approximate time when our ancestors began transitioning from nomadic hunter-gatherers to agricultural domestics.

According to the bone record, however, abstract thought likely started far earlier, as evidenced by the discovery of a burial from 100,000 years ago where grave goods accompanied the body, implying that the deceased's family or tribe understood the concept of an afterlife. But it was approximately 60,000 years later before there is widespread evidence of higher-level thinking and rapid behavioral advances toward cultural modernity. Again, thanks to information provided by osseous relics, these changes apparently came about almost simultaneously in multiple locations. For instance, a cave in Korea has yielded a bone inscribed with a deer from that time. Beginning at about that time in Europe, early humans drew with great artistry and imagination on cave walls; and these findings, along with the discovery of bone rasps and flutes, deer bone necklaces, pendants, and other artifacts of similar vintage, imply that after early humans had met their survival needs, they had some free time to soothe their souls.

In North America, the time line for discoveries of crafting on bone as an indicator of symbolic thinking is much shorter, only a mere 13,000 years. Dated from that time, a bone found in Florida depicts a mammoth carved in profile. From a thousand years later and discovered in Oklahoma came a bison skull painted with a red zigzag. This bone is the oldest known painted object in America.

In the subsequent millennia and in force, humans have carved, painted, and inscribed bone with lifelike figures and abstract shapes. Were these charms, amulets, and fertility fetishes or merely items for amusement and adornment? In many instances, we will never know. The objects do, however, mark the time line of human development when survival was no longer a totally consuming endeavor. Crafted bone found during this time of rapid change range from ordinary (rings, bracelets) to special (studs and long pins for pierced lips, ears, and noses) to completely bizarre (penis ornaments). You can decide whether the latter was a cultural advance or retreat.

Over recent centuries, the craftsmanship of nineteenth-century whalers exemplifies how humans have skillfully painted, engraved, and carved bone. The whalers had time in abundance during their multiyear voyages, along with a ready supply of whalebone and teeth. Herman Melville, in *Moby-Dick* (1851), notes, "But in general, they toil with their jack-knives alone; and, with that almost omnipotent tool of the sailor, they will turn you out anything you please, in the way of a mariner's fancy." Scrimshaw took many forms and today is a highly collectible and often counterfeited folk art, since most of the artists did not sign their work and remain unidentified. Without training, they depicted scenes of their sailing experiences and memories of home on large flat panels of whale jawbone and on the conical surface of whale teeth. The whalers also decorated long narrow strips of bone to give to their sweethearts on return home. These were called busks—stiffeners to fit into front pockets of Victorian-

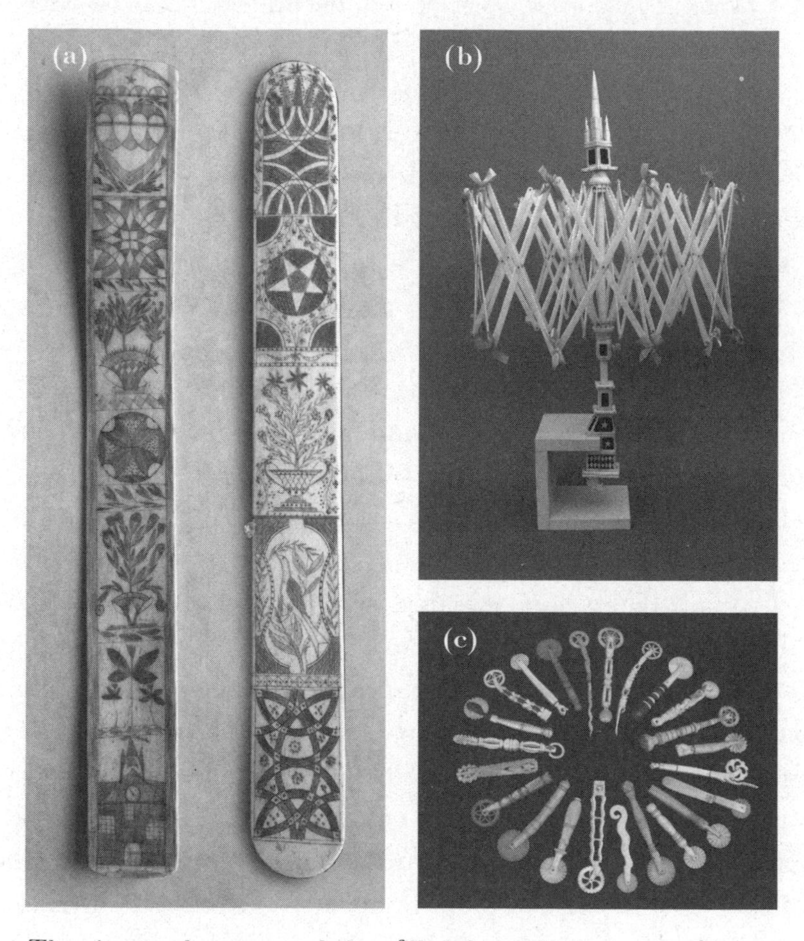


*The nineteenth-century whalers filled their leisure time crafting a wide variety of objects from whalebone. (a) Busks, or corset stays, depicted symbols of love and affection and were worn close to the lucky lady's heart. (b) The swift supported skeins of yarn as they were wound into balls for knitting. (c) Pie crimpers exemplify the whaler-craftsmen's high degree of creativity and skill.* (A) The Nancy Rosin Collection; (B) Courtesy of the New Bedford Whaling Museum; (C) Courtesy of the New Bedford Whaling Museum, Richard Donnelly, photographer.

era corsets. Busks were often sentimentally illustrated and inscribed and were intended to be worn close to the heart to soften the lovers' next time apart. On a less romantic note and maybe hoping for a sweater, whalers turned long narrow strips of bone into swifts, which were sort of collapsible lampshades mounted on a central pivot around which yarn could be wound and then easily removed to form a skein ready for knitting or weaving.

The whalers also transformed bone into pie crimpers, sometimes called jagging wheels, which were used to seal a pie's upper crust—not only dessert pies but also main course ones containing meat or mince. The present-day abundance of these crimpers suggests that many, many piecrusts needed crimping or that the implements were mainly a display of the carver's skill. In favor of the latter, jagging wheels often included delicate filigree and whimsical animals, far beyond the demands of functionality.

Craftspeople have not only used bone to make decorative objects but have also inlaid flat bits of it onto tables, chairs, and chests, thereby turning utilitarian furniture into functional art. The contrast of white bone against the dark surrounding and supportive resin makes for dazzling geometric arrays of mind-boggling intricacy. The craft may have originated in ancient Egypt or in East or South Asia and then spread westward into Europe as trade routes opened up during the Middle Ages. In addition to furniture, high-end saddles, gunstocks, jewel boxes—basically any wooden object—became fair game. In *The Canterbury Tales*, Chaucer describes Sir Topas this way: "His helmet was shiny brass; His armor shone in the sun outside; And on a whale bone saddle he did ride; All other knights he did outclass."

Bone inlay achieves its epitome in the age-old Persian craft of khatamkari, where long thin sticks of camel bone, metal, and wood, each less than a sixteenth of an inch across, are bound and glued into a rod. The artisan stacks and glues the rods together side by side and then saws thin wafers off the end, thin enough to be applied even to curved surfaces of pipes, musical instruments, and other beguiling objects. A square inch of khatam may contain over 400 minuscule

*The ancient Persian craft of khatamkari entailed inlaying sections of thinly sliced rods, consisting of thin strips of camel bone, wood, and wire, that the artisan glued and bound together tightly.* PARIVASH KASHANI.

bits of bone and contrasting materials arranged in dazzling geometric patterns. Jeweler Rene Lalique also incorporated bits of bone into his ornately crafted, art nouveau brooches and pendants.

In aesthetic applications described so far, bone has been the medium with which the artist has delivered a message, perhaps one of respect to a supernatural being, one of love to a spouse, or even one of predilection for well-crimped pies. Using a wide array of non-bone media, artists have also made bone, particularly skulls, the messenger—usually a harbinger of death. The most intriguing and respectful way artists have used bone, however, is to make the bone itself the message: "Look at me, I am beautiful. I come from an animal that also was beautiful." Take, for example, necklaces made by merely stringing unaltered bones together—rattlesnake or fish vertebrae, turtle thighbones, or hundreds of boa constrictor ribs.

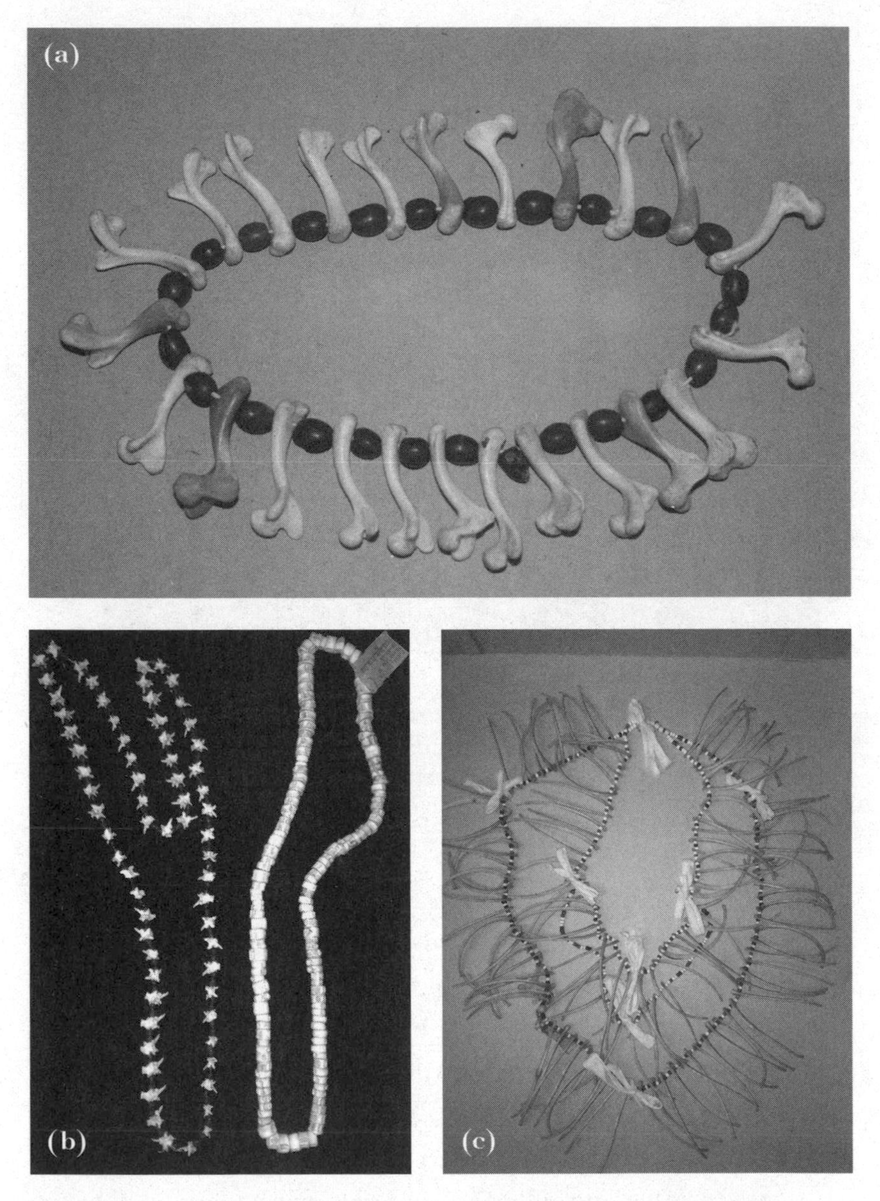


*Bones' smoothly flowing shapes have lent themselves beautifully to satisfy various indigenous people's desire for adornment: (a) turtle thighbones, Oklahoma; (b) rattlesnake vertebrae (left) and fish vertebrae (right), both from Mexico; (c) boa constrictor ribs, South America.* (A) COURTESY OF THE OKLAHOMA HISTORICAL SOCIETY; (B, C) NATURAL HISTORY MUSEUM OF LOS ANGELES COUNTY.

*Antoni Gaudí's extensive 1904 remodel of this previously unremarkable home included the addition of this flowing art nouveau facade, which led local Barcelonans to nickname it the House of Bones.* Casa Batlló, Barcelona.

Further consider that various seacoast communities have erected monumental arches made from whale jawbones. Long and gracefully curved by nature, each one of these skeletal elements simulates the line of a parabola. Architects have recognized this. The Sydney Opera House, the Gateway Arch in St. Louis, and the structural supports of many churches and bridges celebrate this shape. Whales, however, did it first.

Spaniard Antoni Gaudí (1852–1926) not only incorporated parabolic arches into his stunning architecture, but also incorporated bones' flowing contours so extensively on the exterior of his whimsical Casa Batlló that the local Barcelonans nicknamed it the House of Bones.

Then American Georgia O'Keeffe (1887–1986) did with paint

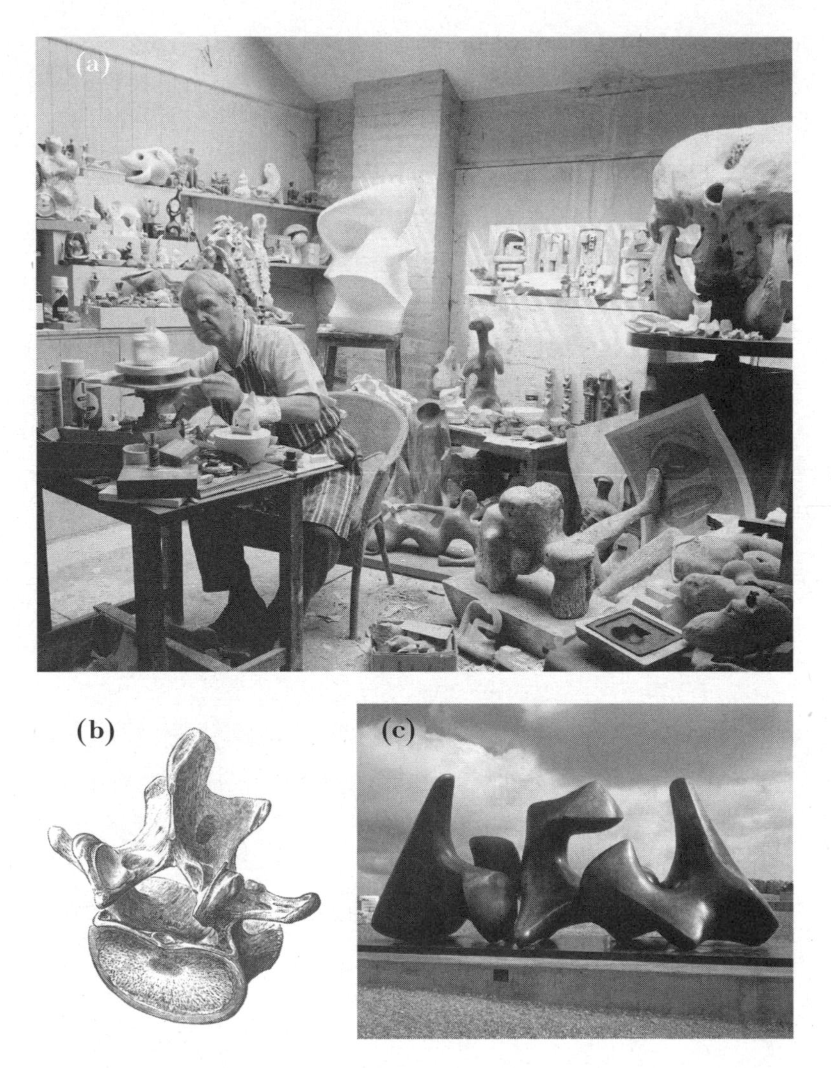


*Renowned British artist Henry Moore picked up bones on his farm and displayed them in his studio to inspire creativity (a). Curving surfaces, merging gracefully into one another, are evident on both a vertebra (b) and in Moore's nature-mimicking, abstract sculptures (c).* (A) REPRODUCED BY PERMISSION OF THE HENRY MOORE FOUNDATION, JOHN HEDGECOE PHOTOGRAPHER; (B) HENRY GRAY, *ANATOMY OF THE HUMAN BODY*, 20TH ED. EDITED BY WARREN LEWIS. (PHILADELPHIA: LEA AND FEBIGER, 1918); (C) REPRODUCED BY PERMISSION OF THE HENRY MOORE FOUNDATION, *THREE PIECE SCULPTURE: VERTEBRAE.*

what Gaudí had done with stone. During her extensive time exploring the hinterlands of New Mexico, she became fond of sun-bleached skulls and pelvises and incorporated them into her paintings. Somewhat abstractly, she ignored the weathered bones' details to capture instead their timeless, graceful curves, causing them to fade from shadow into light and again into shadow.

In my opinion, the highest achievement in using bone for soul-nurturing inspiration goes to British sculptor Henry Moore (1889–1986). He is well known for his semi-abstract bronze forms of reclining women, which grace sculpture parks worldwide. His monumental works often have sweeping curves along with hollow spaces and openings, which some viewers compare to the undulating landscape of Moore's native Yorkshire; but those observers should look more closely at bone. In 1940, Moore was bombed out of his London home and moved to a farmhouse appropriately named Hoglands because it was situated on a former pig farm. On walks, Moore would pick up scattered bones and take them into his studio. He commented that his primary interest was the human form but that he also paid great attention to natural inanimate shapes, including bones, pebbles, and shells. It is apparent that he studied vertebrae closely and was taken by their rounded openings and curved, merging surfaces. They inspired him to create many of his shapes, iconic in their own right and referencing beautiful, timeless bone. Well, almost timeless.

*Chapter 16*

# THE FUTURE OF REVEALED BONE

BONE IN ITS SECOND LIFE—AFTER ITS ORIGINAL owner passes on—has revealed information regarding the most recent 500 million years of Earth's history—the span since bone first appeared. It has also documented 100,000 years of human development and culture. Looking forward, will bone be as revealing in the future as it has been to date? We have barely scratched the surface regarding what old bones have to teach. That some bones fossilize and thereby withstand degradation for millions of years is astounding and wonderful. That some of those fossils eventually reveal themselves and contribute to the telling of Earth's history is spine-tingling.

What fraction of fossils has been unearthed and studied to date? Nobody knows, but it has to be small, likely extremely small. Consider that fossil hunters have to rely on serendipity to discover ancient traces of life. There are not enough paleontologists, research grants, and giant earth movers to systematically and thoroughly dissect the extensive fossil-bearing strata known to exist on every continent, including Antarctica. And in one sense that's good, because such a massive fossil-seeking excavation would destroy most of the objects of its quest in the process. Rather, fossil hunters will have to continue to wait for natural forces and excavations for unrelated purposes to expose long-buried treasures.

Regardless of the means of discovery, the rift between hobbyists and professionals regarding who owns old bones will likely continue. Some suggest protecting fertile fossil fields with national park or World Heritage status. Should this happen, discovery of new species along with more complete skeletons of known species will maximally add to our understanding of Earth's history. As has happily happened with elephant ivory, one day commercial trade in fossils might be banned.

The word *paleontology* was coined 200 years ago, and as the nineteenth century progressed, the discipline developed quickly. But the methods of discovery, recovery, documentation, and preservation of fossilized bones has remained essentially unchanged from the outset: stumble across a bone, meticulously dig it out with dental picks and whisk brooms, measure it, describe it, speculate about its relationship to previous discoveries, and store it in a laboratory drawer or museum. Several aspects of this process are undergoing change, others may follow.

First, take discovery. Paleontologists in the future might sit in their air-conditioned SUVs and fly drones over the bone beds using cameras and computers aided with artificial intelligence to distinguish bone from rock. Incorporating infrared or other extra-human perceptions, the drones could be better than trained eyes. This might sound farfetched, but so did computerized facial recognition several decades back.

How about recovery? We have robotic-assisted surgery, so do not discount the possibility of similarly assisted fossil extraction—once again with the paleontologist supervising from the shade.

Regarding measurement and description of fossils, monumental changes are underway and will accelerate. Paleontologists are naturally loath to break a fossilized bone open, so its interior structure has heretofore been unknown. Conventional X-rays do not penetrate fossils or any stone particularly well (although paleontologists began trying within months after the discovery of X-rays in 1895). Now investigators are coupling an extremely high-intensity X-ray beam, one that would toast living tissue in an instant, with a computed tomographic scanner to image a fossil's inner contours non-destructively. With this new information, mathematical modeling of function is possible, offering insight into the force with which dinosaurs chomped down and whether they balanced on hind limbs alone. High-resolution CT scanning can now digitally distinguish and virtually separate fossils from their stony encasement. Scaled up, this technique might relegate the paleontologist's dental pick to a museum shelf.

Detailed surface features of bones are amenable to laser scanning. This technique works at a distance and the scanner is portable and relatively cheap. Hence, specimens can be scanned in the field. Also, spatial and geometric data can be obtained from large museum specimens that are not easily moved or that will not fit into a CT scanner. Plug this information into a 3-D printer, and presto, out comes a lightweight, high-resolution replica useful for study and instruction anywhere in the world without causing the original any degradation.

Skulls are often flattened and broken during the fossilization process and may be further damaged during retrieval and handling. Accurate anatomical and functional studies, however, rely on analysis of the skull's original form. In the past, investigators have guesstimated that shape after making photographs, sketches, and handmade plastic infills. Now using scanned data, digital reconstruction can remove small breaks, replace missing portions, realign separated segments, and re-expand flattened specimens. To minimize the influence of operator interpretation and (perhaps unrecognized) bias, a computer can dispassionately apply repeatable mathematical techniques automatically and produce a range of reconstructions that can be scrutinized and discussed.

Investigators and institutions are now placing online digitally scanned data of specimens that have been collected over the past 200 years. As this trend continues, researchers can use this "big data" to provide a far more complete picture of large-scale evolutionary patterns regarding not only skeletal forms but also their components' functions and chemical compositions. No longer will our understanding of the history of life be a case of the blind men describing an elephant. Artificial intelligence will likely play a role here, identifying patterns from massive quantities of data where human investigators see only randomness.

These same advances apply to anthropology as well as to paleontology—additional time in the laboratory using high-tech analytical tools to scrutinize existing artifacts. One such example is reexamining the edges of stone tools, now under magnification a thousand times more powerful than the initial naked-eye assessments. Microscopic bits of bone turn up, from which the researchers can extract and study their DNA. More extensive studies along these lines will certainly follow. Another example is the use of DNA recovered from ancient bones to discover heretofore unknown or misunderstood relationships between different species of extinct animals. Previously only gross physical comparisons were possible. Similar detailed chemical comparisons are underway between extinct and present life-forms, such as between Neanderthals and humans.

Can we expect one day that DNA derived from the bones of extinct animals will be used for cloning and reintroduction of ancient species? Ethics and practicality aside, the answer depends on how long the animal has been gone. Yes, researchers are recovering DNA from ancient bones, including from mammoths going back at least 4,000 years, just before they died out. To date, the oldest complete set of genes comes from the cannon bone of a 700,000-year-old extinct species of horse discovered in Yukon Territory permafrost. That is remarkable considering that the DNA molecules begin degenerating at death. Dogma holds that DNA cannot withstand fossilization, but investigators are currently recovering small snippets of DNA from dinosaur fossils a hundred times older than the Yukon horse. Experts are entirely skeptical, however, that DNA strands long enough to carry any genetic information will ever be derived from dinosaur bones. We'll see.

It's a different story, however, for a few proteins. Investigators have successfully extracted and analyzed the amino acid sequence of collagen from 3.5-million-year-old mammalian fossils. Since proteins are coded by DNA, analyzing the protein's amino acid sequence yields an indirect view of the long-gone DNA. Compare it to having a Xerox print of an old-time photograph from which you can surmise what the original negative looked like. In this way, investigators are studying ancient collagen and reverse engineering the small fraction of DNA that controlled collagen production. Since bones and teeth are the only tissues that are preserved through the ages, no proteins from soft tissues are available, so this technique will not produce anywhere near the complete the DNA sequence necessary to recreate an ancient beast.

In far more recent years, Ötzi the Iceman, various mammoths, and the Yukon horse cannon bone have emerged from their frozen graves because of global warming. In the future, other animals will emerge from their deep freezers and enhance our understanding of life on Earth. Will bones of modern humans be around in thousands or tens of thousands of years for additional discoveries and studies? Cremation, presently trending sharply upward in worldwide acceptance, is counterproductive in this regard; and even with burial, eventual disinterment for study and preservation could face more limitations due to cultural sensitivity and respect for the deceased. And yes, it is macabre, yet realistic, to consider that mudslides, avalanches, and volcanic eruptions will provide employment opportunities for future investigators even if study of material from conventional gravesites is further restricted. Despite that dark note, revealed bone's future is exciting and bright for continually enhancing our understanding of Earth's history.

What about the future for revealed bone's other capacity—to record human culture? First, a word about what its ongoing role will not be. Except for hobbyists, bone's long run—in some instances ranging into tens of thousands of years—as the go-to material for making needles, flutes, fishhooks, louse killers, spoons, combs, and dice is over. May museums carefully preserve and display these icons of human culture in perpetuity. And in this regard, institutions are rapidly placing their complete collections online to facilitate worldwide appreciation and study. To see what I mean, find your favorite anthropology, paleontology, natural history, science, medical, veterinary, musical instrument, decorative art, or fine art museum on the Internet, click on *search collection*, type in *bone*, and enjoy. I frequently do and learn so much. As a side benefit, electronic access to the collections also avoids repeatedly handling and inevitably degrading the objects.

Even though there will be no recently minted bone buttons and arrow points for discovery and study in the future, bone will take on new roles as cultural markers with growing frequency. Enterprising architects, engineers, and product designers will increasingly use bone's beautiful contours as inspiration. These forms, which nature has perfected over 500 million years, reflect living bone's function—to efficiently support its owner. Computer design and expanding 3-D printing capabilities will accelerate this trend, since these tools can readily replicate life-forms that were previously impossible to emulate using conventional manufacturing processes. The sculptors and painters who have already extolled bone's beauty are setting the bar for others to raise. Far into the future and in so many ways, the beauty, efficiency, and timelessness of bone will make themselves regularly apparent, and that deserves awe and celebration.

◆◆◆◆

THESE ARE IMMORTAL bones' great stories. Protein chains linked to one another and coated with calcium crystals form a unique substance that supports advanced animal forms. When alive, with few exceptions, bone remains concealed. In its second life, bone reveals much about Earth's history and human activity. It is both an inheritance and a legacy. Bone is the world's best building material. It always has been. It always will be.

# *Acknowledgments*

Agent Gillian MacKenzie immediately and enthusiastically endorsed my idea for a comprehensive book about bone. Together with Allison Devereux, they molded the proposal and guided me to publisher W. W. Norton, where editor Quynh Do and her talented team formed the work into the final polished product all authors hope for. Also essential to the book was the work of graphic artist Renee Pulvee, who ensured that all of the images were clear and elegant. Early on, Peter O'Neill, business and mentor coach at SCORE Los Angeles, provided invaluable insight on improving the social media presence for my blog and awareness of the book's eventuality.

Unseen to the reader are the contributions made by the thousands of patients who over my career have entrusted the care of their musculoskeletal problems to me. Our interactions have deeply enriched my understanding and respect for bone, which I joyfully share. Second in numbers comes the multitude of medical students, residents, and fellows who have been my teachers, although ostensibly my students. Their energy and enthusiasm have made learning a stimulating joint venture. Then my bona fide teachers, from junior high school on, deserve recognition for their love of science and curiosity that they instilled in me.

A book of this breadth, spanning information from protein chemistry to popular culture, relies on the expertise of a broad range of specialists. I am deeply indebted to the colleagues, friends, and sages (often one in the same) for their wisdom, suggestions, critique, and enthusiastic support. I would like to particularly recognize and thank Stanley Chernikoff, professor of geology, University of Washington; KT Hajeian, anthropology collections manager, Natural History Museum of Los Angeles County; Russell Johnson, curator, UCLA Library Special Collections; J. Michael Kabo, professor emeritus of mechanical engineering, California State University, Northridge; David Kronen, Managing Director, Bone Clones, Inc; Natalie Langley, senior associate consultant, anatomy, Mayo Clinic School of Medicine; Diana Mansfield, owner, The Bone Room; Mark Procknik, librarian, New Bedford Whaling Museum; David A. Rubin, MD, retired professor of radiology, Washington University St. Louis, and consultant radiologist, St. Louis Zoo; J. Chris Sagebiel, collections manager, Texas Vertebrate Paleontology Collections; William F. Simpson, head of geological collections, Field Museum of Natural History; Julie K. Stein, executive director, Burke Museum; Daniel C. Swan, professor of anthropology, University of Oklahoma; Michele Tabencki, osteologist, Bone Clones, Inc; and Jay Villemarette, owner, Museum of Osteology.

This book would have been impossible without the Internet and instant access to resources worldwide. Online images and descriptions of tens of thousands of bone objects archived in over a hundred anthropological, paleontological, natural history, science, musical instrument, whaling, and fine art museums proved to be invaluable study material. I am grateful to these museums as well as to the private collectors who provided images that I have included. Also, comprehensive electronic access to books and journals, both contemporary and ancient, was joyful and enlightening, and for that I am indebted to the UCLA Libraries.

# *Bibliography*

## *Chapter 1:* BONE'S UNIQUE COMPOSITION AND VARIED STRUCTURE

Alexander, R. McNeill. *Bones. The Unity of Form and Function*. New York: Nevraumont, 1994.

Alexander, R. McNeill. *Human Bones. A Scientific and Pictorial Investigation*. New York: Nevraumont, 2005.

Ashby, Michael. *Materials Selection in Mechanical Design*. 4th ed. Burlington, MA: Butterworth-Heinemann, 2011.

Associated Press. "Walrus Penis Sells for $8,000 at Beverly Hills Action [sic]." Accessed September 21, 2019. https://web.archive.org/web/20071106050910/http:/www.sfgate.com/cgi-bin/article.cgi?f=/n/a/2007/08/26/state/n154935D40.DTL

Burt, William. *Bacula of North American Mammals*. Ann Arbor: University of Michigan Press, 1960.

Currey, John. *Bones, Structure and Mechanics*. Princeton, NJ: Princeton University Press, 2006.

Duncker, Hans-Rainer. "Structure of the Avian Respiratory Tract." *Respiration Physiology* 22, no. 1–2 (1974): 1–19.

Farmer, C. G. "On the Origin of Avian Air Sacs." *Respiratory Physiology and Neurobiology* 154, no. 1–2 (2006): 89–106.

Goodsir, John. "The Structure and Economy of Bone." In *Classics of Orthopaedics*, 79–81. Edited by Edgar Bick. Philadelphia: Lippincott, 1976.

"A History of the Skeleton." Accessed September 21, 2019. https://web.stanford.edu/class/history13/earlysciencelab/body/skeletonpages/skeleton.html

Jellison, W. L. "A Suggested Homolog of the Os Penis or Baculum of Mammals." *Journal of Mammalogy* 26, no. 2 (1945): 146–47.

Johnson, Robert Jr. "A Physiological Study of the Blood Supply of the Diaphysis." *Journal of Bone and Joint Surgery* 9, no. 1 (1927): 153–84.

Lambe, Lawrence. "The Cretaceous Theropodous Dinosaur *Gorgosaurus*." *Canada Department of Mines Geological Survey Memoir* 100, no. 83 Geological Series (1917): 1–84.

Lambe, Lawrence. "On the Fore-Limb of a Carnivorous Dinosaur from the Belly River Formation of Alberta, and a New Genus of Ceratopsia from the Same Horizon, with Remarks on the Integument of Some Cretaceous Herbivorous Dinosaurs." *Ottawa Naturalist* 27, no. 10 (1914): 129–35.

Layne, James. "The Os Clitoridis of Some North American Sciuridae." *Journal of Mammalogy* 35, no. 3 (1954): 357–66.

O'Connor, Jingmai, Xiao-Ting Zheng, Xiao-Li Wang, Xiao-Mei Zhang, and Zhou Zhong-He. "The Gastral Basket in Basal Birds and Their Close Relatives: Size and Possible Function." *Vertebrata PalAsiatica* 53, no. 2 (2015): 133–52.

Parry, David, and John Squire. *Fibrous Proteins: Structures and Mechanisms*. Cham, Switzerland: Springer, 2017.

Ramm, Steven. "Sexual Selection and Genital Evolution in Mammals: A Phylogenetic Analysis of Baculum Length." *American Naturalist* 169, no. 3 (2007): 360–69.

Roycroft, Patrick D, and Martine Cuypers. "The Etymology of the Mineral Name 'Apatite': A Clarification." *Irish Journal of Earth Sciences* 33 (2015): 71–75.

Schmitz, Lars, and Ryosuke Motani. "Nocturnality in Dinosaurs Inferred from Scleral Ring and Orbit Morphology." *Science* 332, no. 6030 (2011): 705–8.

Singer, Charles. "Galen's Elementary Course on Bones." *Proceedings of the Royal Society of Medicine* 45, no. 11 (1952): 767–76.

Steele, Gentry, and Claud Bramblett. *The Anatomy and Biology of the Human Skeleton*. College Station: Texas A and M University Press, 2008.

Weishampel, D. B. "Acoustic Analysis of Vocalization of Lambeosaurine Dinosaurs (Reptilia: Ornithischia)." *Paleobiology* 7, no. 2 (1981): 252–61.

Yamashita, Momo, Takuya Konisi, and Tamaki Sato: "Sclerotic Rings in Mosasaurs (Squamata: Mosasauridae): Structures and Taxonomic Diversity." *PLoS One* (February 18, 2015). Accessed September 21, 2019. http://dx.doi.org/10.1371/journal.pone.0117079.

Young, Barbara, and John Heath. *Wheater's Functional Histology*. Edinburgh: Churchill Livingstone, 2000.

## *Chapter 2:* BONE'S LIFE AND RELATIVES

Alexander, Robert. *Bones. The Utility of Form and Function*. New York: Nevraumont Publishing Company, 1994.

Alexander, Robert. *Human Bones, A Scientific and Pictorial Investigation*. New York: Pearson Education, 2005.

Blount, Walter, and George Clarke. "Control of Bone Growth by Epiphyseal Stapling." In *Classics of Orthopaedics*, 371–84. Edited by Edgar Bick. Philadelphia: Lippincott, 1976.

Bronikowski, Anne. "The Evolution of Aging Phenotypes in Snakes: A Review and Synthesis with New Data." *Age* 30, no. 2–3 (2008): 169–76.

Dobson, Jessie. "Pioneers of Osteogeny: Clopton Havers." *Journal of Bone and Joint Surgery* 34 B, no. 1 (1952) 702–7.

Dykens, Margaret, and Lynett Gillette. "Giant Sloth." Accessed September 21, 2019. https://www.sdnhm.org/exhibitions/fossil-mysteries/fossil-field-guide-a-z/giant-sloth/

Feagans, Carl. "Artificial Cranial Modification in the Ancient World." Accessed September 22, 2019. http://www.academia.edu/278283/

Foerster, Brien. *Elongated Skulls of Peru and Bolivia: The Path of Viracocha*. San Bernadino: Brien Foerster, 2015.

Halliday, T. R., and P. A. Verrell. "Body Size and Age in Amphibians and Reptiles." *Journal of Herpetology* 22, no. 3 (1988): 253–65.

Hariharan, Iswar, David Wake, and Marvalee Wake. "Indeterminate Growth: Could It Represent the Ancestral Condition?" *Cold Spring Harbor Perspectives in Biology* 8, no. 2 (2016): 1–17.

Jones, H. H., J. D. Priest, W. C. Hayes, C. C. Tichenor, and D. A. Nagel. "Humeral Hypertrophy in Response to Exercise." *Journal of Bone and Joint Surgery, American* 59, no. 2 (1977): 204–8.

Kontulainen, Saiji, Harri, Sievanen, Pekka Kannus, Matti Pasanen, and Vuori Ilkka. "Effect of Long-Term Impact-Loading on Mass, Size, and Estimated Strength of Humerus and Radius of Female Racquet-Sports Players: A Peripheral Quantitative Computed Tomography Study between Young and Old Starters and Controls." *Journal of Bone Mineral Research* 18, no. 2 (2003): 352–59.

Madsen, Thomas, and Richard Shine. "Silver Spoons and Snake Body Sizes: Prey Availability Early in Life Influences Long-Term Growth Rates of Free-Ranging Pythons." *Journal of Animal Ecology* 69, no. 6 (2000): 952–58.

McLean, Franklin, and A. Baird Hastings. "The State of Calcium in the Fluids of the Body." In *Classics of Orthopaedics*, 292–315. Edited by Edgar Bick. Philadelphia: Lippincott, 1976.

Reynolds, Gretchen. "How Our Bones Might Help Keep Our Weight in Check." *New York Times*, January 17, 2018. Accessed September 21, 2019. https://www.nytimes.com/2018/01/17/well/move/how-our-bones-might-help-keep-our-weight-stable.html

Shine, Richard, and Eric Charnov. "Patterns of Survival, Growth, and Maturation in Snakes and Lizards." *American Naturalist* 139, no. 6 (1992): 1257–69.

Tiesler, Vera. "Studying Cranial Vault Modifications in Ancient Moamerica." *Journal of Anthropological Sciences* 90 (2012): 33–58.

Trinkaus, Erik. "Artificial Cranial Deformation in the Shanidar 1 and 5 Neandertals." *Current Anthropology* 23, no. 2 (1982): 198–99.

### *Chapter 3:* WHEN BONES BREAK

Amstutz, Harlan, Eric Johnson, Gerald Finerman, Roy Meals, John Moreland, William Kim, and Marshall Urist. "New Advances in Bone Research." *Western Journal of Medicine* 141, no. 1 (1984): 71–87.

Court-Brown, Charles, James Heckman, Margaret McQueen, William Ricci, Paul Tornetta III, and Michael McKee, eds. *Rockwood and Green's Fractures in Adults.* 8th ed. Philadelphia: Lippincott Williams & Wilkins/Wolters Kluwer Health, 2015.

Flynn, John, David Skaggs, and Peter Waters, eds. *Rockwood and Wilkins' Fractures in Children.* 8th ed. Philadelphia: Lippincott Williams & Wilkins/Wolters Kluwer Health, 2015.

Jones, Robert. "An Orthopaedic View of the Treatment of Fractures." In *Classics of Orthopaedics*, 348–60. Edited by Edgar Bick. Philadelphia: Lippincott, 1976.

Miller, Timothy, and Christopher Kaeding, eds. *Stress Fractures in Athletes: Diagnosis and Management.* Cham, Switzerland: Springer, 2014.

Peltier, Leonard. *Fractures. A History and Iconography of Their Treatment.* San Francisco: Norman Publishing, 1990.

Thomas, Hugh. "Diseases of the Hip, Knee and Ankle Joint with Their Deformities Treated by a New and Efficient Method." In *Classics of Orthopaedics*, 469–74. Edited by Edgar Bick. Philadelphia: Lippincott, 1976.

### *Chapter 4:* BONE'S OTHER FAILINGS AND WHO CAN HELP

"Abaloparatide (Tymlos) for Postmenopausal Osteoporosis." *The Medical Letter on Drugs and Therapeutics* 59, issue 1523 (2017): 97–98.

Aegeter, Ernest, and John Kirkpatrick Jr. *Orthopedic Diseases: Physiology, Pathology, Radiology.* Philadelphia: W. B. Saunders, 1975.

Blount, Walter, and George Clarke. "Control of Bone Growth by Epiphyseal Stapling. A Preliminary Report." *Journal of Bone and Joint Surgery* 31A, no. 3 (1949): 464–78.

Doherty, Alison, Cameron Ghalambor, and Seth Donahue. "Evolutionary Physiology of Bone: Bone Metabolism in Changing Environments." *Physiology* 30, no. 1 (2015): 17–29.

Doherty, Alison, Danielle Roteliuk, Sara Gookin, Ashley McGrew, Carolyn Broccardo, Keith Condon, Jessica Prenni, et al. "Exploring the Bone Proteome to Help Explain Altered Bone Remodeling and Preservation of Bone Architecture and Strength in Hibernating Marmots." *Physiological and Biochemical Zoology* 89, no. 5 (2016): 364–76.

Everett, E. T. "Fluoride's Effects on the Formation of Teeth and Bones, and the Influence of Genetics." *Journal of Dental Research* 90, no. 5 (2011): 552–60.

Freese, Barbara. *Coal: A Human History*. New York: Perseus, 2003.

Hillier S., H. Inskip, D. Coggon, and C. Cooper. "Water Fluoridation and Osteoporotic Fracture." *Community Dental Health*, Supplement 2 (1996): 63–68.

Kanavel, Allen. *Infections of the Hand. A Guide to the Surgical Treatment of Acute and Chronic Suppurative Processes in the Fingers, Hand, and Forearm*. Philadelphia: Lea and Febiger, 1912.

Kohlstadt, Ingrid, and Kenneth Cintron, eds. *Metabolic Therapies in Orthopedics*. 2nd ed. Boca Raton, FL: CRC Press, an imprint of Taylor and Francis Group, 2019.

McGee-Lawrence, Meghan, Patricia Buckendahl, Caren Carpenter, Kim Henriksen, Michael Vaughan, and Seth Donahue. "Suppressed Bone Remodeling in Black Bears Conserves Energy and Bone Mass during Hibernation." *Journal of Experimental Biology* 218 (2015): 2067–74.

Meals, Roy. *The Hand Owner's Manual. A Hand Surgeon's Thirty-Year Collection of Important Information and Fascinating Facts*. College Station, TX: Virtualbook worm.com, 2008.

Meals, Roy, and Scott Mitchell. *One Hundred Orthopedic Conditions Every Doctor Should Understand*. 2nd ed. St. Louis, MO: Quality Medical Publishing, 2006.

Møller, P. Flemming, and Sk V. Gudjonsson. "Massive Fluorosis of Bones and Ligaments." *Acta Radiologica* 13, no. 3–4 (1932): 269–94.

Olson, Steven, and Farshi Guilak, eds. *Post-traumatic Arthritis: Pathogenesis, Diagnosis and Management*. New York: Springer, 2015.

Pandya, Nirav, Keith Baldwin, Atul Kamath, Dennis Wenger, and Harish Hosalkar. "Unexplained Fractures: Child Abuse or Bone Disease? A Systematic Review." *Clinical Orthopaedics and Related Research* 469, no. 3 (2011): 805–12.

Paschos, Nikolaos, and George Bentley, eds. *General Orthopaedics and Basic Science*. Cham, Switzerland: Springer, 2019.

Petrone, Pierpaolo, Michele Giordano, Stefano Giustino, and Fabio Guarino. "Enduring Fluoride Health Hazard for the Vesuvius Area Population: The Case of AD 79 Herculaneum." *PLoS One* (June 16, 2011). https://doi.org/10.1371/journal.pone.0021085

Phipps, Kathy, Eric Orwoll, Jill Mason, and Jane Cauley. "Community Water Fluoridation, Bone Mineral Density, and Fractures: Prospective Study of Effects in Older Women." *British Medical Journal* 321, no. 7255 (2000): 860–64.

Picci, Piero, Marco Manfrini, Nicola Fabbi, Marco Gammbarotti, and Daniel Vanel, eds. *Atlas of Musculoskeletal Tumors and Tumorlike Lesions: The Rizolli Case Archive*. Cham, Switzerland: Springer, 2015.

Prada, Diddier, Elena Colicino, Antonella Zanobetti, Joel Schwartz, Nicholas Dagincourt, Shona Fang, Itai Kloog, et al. "Association of Air Particulate Pollution with Bone Loss over Time and Bone Fracture Risk: Analysis of Data from Two Independent Studies." *Lancet Planetary Health* 1, no. 8 (2017): PE 337–E347.

Rozbruch, Robert, and Reggie Hamdy, eds. *Limb Lengthening and Reconstruction Surgery Case Atlas*. Cham, Switzerland: Springer, 2015.

Shapiro, Frederic. *Pediatric Orthopedic Deformities. Volume 1, Pathobiology and Treatment of Dysplasias, Physeal Fractures, Length Discrepancies, and Epiphyseal and Joint Disorders*. Cham, Switzerland: Springer, 2015.

Staheli, Lynn. *Fundamentals of Pediatric Orthopedics*. Philadelphia: Wolters Kluwer, 2016.

Taylor, Robert Tunstall. *Orthopaedic Surgery for Students and General Practitioners: Preliminary Considerations and Diseases of the Spine; 114 Original Illustrations*. Baltimore: Williams & Wilkins, 1907.

Whitney, William. Bulletin of the Warren Anatomical Museum, no. 1, *Pathological Anatomy, Bones, Joints, Synovial Membranes, Tendons*. Boston: Harvard Medical School, 1910.

Wojda, Samantha, Richard Gridley, Meghan McGee-Lawrence, Thomas Drummer, Ann Hess, Franziska Kohl, Brian Barnes, and Seth Donahue. "Arctic Ground Squirrels Limit Bone Loss during the Prolonged Physical Inactivity Associated with Hibernation." *Physiological and Biochemical Zoology* 89, no. 1 (2016): 72–80.

## *Chapter 5:* BONE SURGERY THROUGH THE AGES

Andry, Nicholas. *Orthopédie*. Paris: La Veuve Alix, 1741.

"A. T. Still: A Profile of the Founder of Osteopathy." Accessed September 25, 2019. https://web.archive.org/web/20120426232748/http://www.osteohome.com/SubPages/Still.html

Chambers, Caitlin, Stephanie Ihnow, Emily Monroe, and Linda Suleiman. "Women in Orthopaedic Surgery: Population Trends in Trainees and Practicing Surgeons." *Journal of Bone and Joint Surgery, American* 100, no. 17 (2018): e116.

Duncan, Gregory, and Roy Meals. "One Hundred Years of Automobile-Induced Orthopaedic Injuries." *Orthopedics* 18, no. 2 (1995): 165–70.

Dydra, Laura. "8 Orthopedic Surgeons Who Are Famous Outside of Orthopedics." Accessed October 3, 2019. https://www.beckersspine.com/spine-lists/item/24430–8-orthopedic-surgeons-who-are-famous-outside-of-orthopedics

Freedman, Eric, Marc Safran, and Roy Meals. "Automobile Air Bag-related Upper Extremity Injuries. A Report of Three Cases." *Journal of Trauma* 38, no. 4 (1995): 577–81.

Harness, Neil, and Roy Meals. "The History of Fracture Fixation of the Hand and Wrist." *Clinical Orthopaedics and Related Research* 445 (2006): 19–29.

Jones, Robert. "An Orthopaedic View of the Treatment of Fractures." *Clinical Orthopaedics and Related Research* 75 (March–April 1971): 4–16.

LeVay, David. *The History of Orthopaedics*. Carnforth, UK: Parthenon, 1990.

Lyons, Albert, and R. Joseph Petrucelli II. *Medicine. An Illustrated History.* New York: Harry N. Abrams, 1978.

Manjo, Guido. *The Healing Hand: Man and Wound in the Ancient World.* Cambridge, MA: Harvard University Press, 1975.

Meals, Clifton, and Roy Meals. "Hand Fractures: A Review of Current Treatment Strategies." *Journal of Hand Surgery, American* 38, no. 5 (2013): 1021–31.

Meals, Roy. "Surgical Teaching vs. Surgical Learning." *Loyola University Orthopaedic Journal* 2 (1993): 35–38.

Meals, Roy. "Teaching Clinical Judgement. Teaching the Choice of Surgical Procedures in the Treatment of Arthritis of the Hip." *British Journal of Medical Education* 7, no. 2 (1973): 100–102.

Meals, Roy, and Christof Meuli. "Carpenter's Nails, Phonograph Needles, Piano Wire and Safety Pins: The History of Operative Fixation of Metacarpal and Phalangeal Fractures." *Journal of Hand Surgery, American* 10, no. 1 (1985): 144–50.

Meals, Roy, and Hugh Watts. "Clinicians Teaching Orthopaedics: Effective Strategies." *Instructional Course Lectures* 47 (1997): 583–94.

Melchior, Julie, and Roy Meals. "The Journal Club and Its Role in Hand Surgery Education." *Journal of Hand Surgery, American* 23: no. 6 (1998): 972–76.

Paré, Ambroise. *The Apoligie and Treatise of Ambroise Paré Containing the Voyages Made into Divers Places with Many of His Writings Upon Surgery.* Edited by Geoffrey Keynes. New York: Dover Publications, 1968.

Peltier, Leonard. *Orthopedics. A History and Iconography.* San Francisco: Norman Publishing, 1993.

Singer, Charles. "Galen's Elementary Course on Bones." *Proceedings of the Royal Society of Medicine* 45, no. 11 (1952): 767–76.

Smith, G. Elliot. "The Most Ancient Splints." *British Medical Journal* 1, no. 2465 (March 28, 1908): 732–34.

Thomas, Hugh. *Diseases of the Hip, Knee and Ankle Joint with Their Deformities Treated by a New and Efficient Method.* 3rd ed. London: H. K. Lewis, 1878.

Yang, Paul, and Roy Meals. "How to Establish an Interactive eConference and eJournal Club." *Journal of Hand Surgery, American* 39, no. 1 (2014): 129–33.

## *Chapter 6:* SIX ORTHOPEDIC GIANTS

American Academy of Orthopaedic Surgeons. "Arresting Development. Paul Harrington, MD." Accessed December 1, 2019. http://www.aaos75.org/stories/physician_story.htm?id=8

Bagnoli, Gianfanco. *The Ilizarov Method.* Philadelphia: B. C. Decker, 1990.

Born, Christopher, Tyler Pidgeon, and Gilbert Taglang. "75 Years of Contemporary Intramedullary Nailing." *Journal of Orthopaedic Trauma* 28, Supplement 8 (2014): S1–S2.

Brand, Richard. "Marshall R. Urist, 1914–2001." *Clinical Orthopaedics and Related Research* 467, no. 12 (2009): 3049–50.

Charnley, John. "Arthroplasty of the Hip: A New Operation." In *Classics of Orthopaedics*, 447–51. Edited by Edgar Bick. Philadelphia: Lippincott, 1976.

Charnley, John. *Low Friction Arthroplasty of the Hip: Theory and Practice*. Berlin: Springer-Verlag, 1979.

Douglas, Martin. "Dr. Jacquelin Perry, Surgeon Who Aided Polio Victims, Dies at 94." *New York Times*. Accessed September 23, 2019. https://www.nytimes.com/2013/03/24/health/dr-jacquelin-perry-who-aided-polio-victims-dies-at-94.html

Elliot, Carol, and Joan Headley. "Paul Randall Harrington, MD." Polio Place. Accessed September 24, 2019. https://www.polioplace.org/people/paul-r-harrington-md

Festino, Jennifer. "Giants in Orthopaedic Surgery: Jacquelin Perry MD, DSc (Hon)." *Clinical Orthopaedics and Related Research* 472, no. 3 (2014): 796–801.

Finerman, Gerald. "Marshall R. Urist, MD, 1914–2001." *Journal of Bone and Joint Surgery, American* 83, no. 10 (2001): 1611.

Huggins, Charles. "The Formation of Bone under the Influence of Epithelium of the Urinary Tract." *Archives of Surgery*, 22, no. 3 (1931): 377–408.

Ilizarov, Svetlana. "The Ilizarov Method: History and Scope." In *Limb Lengthening and Reconstructive Surgery*. Edited by S. Robert Rozbruch and Svetlana Ilizarov. Boca Raton: CRC Press, 2007.

Jackson, John: "Father of the Modern Hip Replacement: Professor Sir John Charnley (1911–82)." *Journal of Medical Biography* 19, no. 4 (2011): 151–56.

Jackson, Robert. "A History of Arthroscopy." *Arthroscopy* 26, no. 1 (2010): 91–103.

Lindholm, Ralf. *The Bone-Nailing Surgeon G. B. G. Kuentscher and the Finns*. Oulu, Finland: University of Oulu, 1982.

Özyener, Fadil. "Gait Analysis: Normal and Pathological Function." *Journal of Sports Science and Medicine* 9, no. 2 (2010): 353.

Peltier, Leonard. *Orthopedics. A History and Iconography*. San Francisco: Norman Publishing, 1993.

Perry, Jacquelin. *Gait Analysis. Normal and Pathological Function*. Thorofare NJ: SLACK, 1992.

Reynolds L. A., and E. M. Tansey, eds. "Early Development of Total Hip Replacement." *Wellcome Witnesses to Twentieth Century Medicine*, 29 (2007): 1–167.

Ridlon, John, Hugh Thomas, and Robert Jones. *Lectures on Orthopedic Surgery*. Philadelphia: E. Stern, 1899.

Saxon, Wolfgang. "Dr. Marshall Raymond Urist, 85; Identified Bone-Mending Protein." *New York Times*. Accessed September 24, 2019. https://www.nytimes.com/2001/02/12/us/dr-marshall-raymond-urist-85-identified-bone-mending-protein.html

"Spines of Steel." *Time* 76, no. 20 (1960): 56.

Watts, Geoff. "Jacquelin Perry." *Lancet* 381, no. 9876 (2013): 1454.

Whitman, Royal. *A Treatise on Orthopedic Surgery*. Philadelphia: Lea Brothers, 1903.

### *Chapter 7:* ORTHOPEDIC INNOVATIONS

Ackman, J., H. Altiok, A. Flanagan, M. Peer, A. Graf, J. Krzak, S. Hassani, et al. "Long-Term Follow-Up of Van Nes Rotationplasty in Patients with Congenital Proximal Focal Femoral Deficiency." *Bone and Joint Journal* 95B, no. 2 (2013): 192–98.

Bong, Matthew, Kenneth Koval, and Kenneth Egol. "The History of Intramedullary Nailing." *Bulletin of the NYU Hospital for Joint Diseases* 64, no. 3–4 (2006): 94–97.

Çakmak, Mehmet, Cengiz Şen, Levent Erlap Halik Balci, and Melih Civan. *Basic Techniques for Extremity Reconstruction: External Fixator Applications According to Ilizarov Principles*. Cham, Switzerland: Springer, 2018.

Dahman, Yaser. *Biomaterials Science and Technology: Fundamentals and Developments*. Boca Raton, FL: CRC Press, 2019.

Degryse, Patrick, David De Muynk, Steve Delporte, Sara Boyen, Laure Jadoul, Joan De Winne, Tatiana Ivaneanu, and Frank Vanhaecke. "Strontium Isotope Analysis as an Experimental Auxiliary Technique in Forensic Identification of Human Remains." *Analytical Methods* 4, no. 9 (2012): 2674–79.

Hung, Ben, Bilal Naved, Ethan Nyberg, Miguel Dias, Christina Holmes, Jennifer Elisseeff, Amir Dorafshar, and Warren Grayson. "Three-Dimensional Printing of Bone Extracellular Matrix for Craniofacial Regeneration." *ACS Biomaterials Science & Engineering* 2, no. 10 (2016): 1806–16.

Li, Bingyun, and Thomas Webster: *Orthopedic Biomaterials: Advances and Applications*. Cham, Switzerland: Springer, 2017.

Meals, Roy. "Thumb Reconstruction Following Major Loss. A Review of Treatment Alternatives." *Journal of Trauma* 28, no. 6 (1988): 746–50.

National Heart, Lung, and Blood Institute. "Bone Marrow Transplantation." Accessed September 24, 2019. https://medlineplus.gov/bonemarrowtransplantation.html

Petersen, Traci. "Facts about Strontium." Live Science. Accessed September 24, 2019. https://www.livescience.com/34522-strontium.html

Schoch, Bradley, Michael Hast, Samir Mehta, and Surena Namdari. "Not All Polyaxial Locking Screw Technologies Are Created Equal: A Systematic Review of the Literature." *Journal of Bone and Joint Surgery Reviews* 6, no. 1 (2018): e6.

Wendell, Emely. "Why Strontium Is Not Advised for Bone Health." American Bone Health. Accessed December 1, 2019. https://americanbonehealth.org/medications-bone-health/why-strontium-is-not-advised-for-bone-health/

Wheeless, Cliford III. "Stress Shielding from Femoral Components." Accessed September 24, 2019. http://www.wheelessonline.com/ortho/stress_shielding_from_femoral_components

Wilson, June, and Larry Hench, eds. *Clinical Performance of Skeletal Prostheses*. Boca Raton, FL: Chapman and Hall, 1996.

### *Chapter 8:* PICTURING BONE

"Airport X Ray Scanners." Accessed September 22, 2019. https://www.radiationanswers.org/radiation-blog/airport_xray_scanners.html

Armstrong, April, and Mark Hubbard. *Essentials of Musculoskeletal Care.* Enhanced 5th ed. Burlington, MA: American Academy of Orthopaedic Surgeons, 2018.

Bradley, William. "History of Medical Imaging." *Proceedings of the American Philosophical Society* 152, no. 3 (2008): 349–61.

Chandra, Ramesh, and Arman Rahmin. *Nuclear Medicine Physics. The Basics.* 8th ed. Philadelphia: Lippincott Williams & Wilkins, 2017.

Cheselden, William. *Osteographia, or the Anatomy of the Bones.* London: W Bowyer, 1733.

Cope, Zachary. *William Cheselden 1688–1752.* Edinburgh: E & S Livingstone, 1953.

DeLint, J. G. *Atlas of the History of Medicine.* New York: Hoeber, 1926.

Elgazzar, Abdelhamid. *Orthopedic Nuclear Medicine.* 2nd ed. Berlin: Springer Verlag, 2004.

Glazar, Ed. "How Many Bones Did Evel Knievel Break?" Magic Valley. Accessed October 3, 2019. https://magicvalley.com/news/local/how-many-bones-did-evel-knievel-break/article_a64def32–2d63–11e4-bfc7–0019bb2963f4.html

Greenspan, Adam. *Orthopedic Imaging: A Practical Approach.* 6th ed. Philadelphia: Wolters Kluwer, 2015.

Helms, Clyde. *Fundamentals of Skeletal Radiology.* 5th ed. Amsterdam: Elsevier, 2019.

Illés, Tamás, and Szabolcs Somoskeöy. "The EOS™ Imaging System and Its Uses in Daily Orthopaedic Practice." *International Orthopaedics* 36, no. 7 (2012): 1325–31.

Lin-Watson, TerriAnn. *Radiographic Pathology.* 2nd ed. Philadelphia: Lippincott Williams & Wilkins: 2014.

Love, Charito, Anabella Din, Maria Tomas, Tomy Kalapparambath, and Christopher Palestro. "Radionuclide Bone Imaging: An Illustrative Review." *Radiographics* 23, no. 2 (2003): 341–58.

Malakhova, Olga. "Nikolay Ivanovich Pirogoff (1810–1881)." *Clinical Anatomy* 17, no. 5 (2004): 369–72.

Meals, Roy, and J. Michael Kabo. "Computerized Anatomy Instruction." *Clinics in Plastic Surgery* 13, no. 3 (1986): 379–88.

Meals, Roy, and Leanne Seeger. *An Atlas of Forearm and Hand Cross-sectional Anatomy with Computed Tomography and Resonance Imaging Correlation.* London: Martin Dunitz, 1991.

Neher, Allister. "The Truth about Our Bones: William Cheselden's *Osteographia.*" *Medical History* 54, no. 4 (2010): 517–28.

Peterson, Jeffrey. *Berquist's Musculoskeletal Imaging Companion.* 3rd ed. Phladelphia: Lippincott Williams & Wilkins, 2017.

Pirogov, Nikolai. *An Illustrated Topographic Anatomy of Saw Cuts Made in Three Dimensions across the Frozen Human Body (Atlas, Part 4)* (*Anatome topographica:*

*sectionibus per corpus humanum congelatum: triplici directione ductis illustrata*). St. Petersburg: Typis Jacobi Trey, 1852–1859.

"Radiation Doses in X-Ray and CT Exams." Accessed October 1, 2019. https://www.radiologyinfo.org/en/pdf/safety-xray.pdf

"Radiation Risk from Medical Imaging." Accessed October 1, 2019. https://www.health.harvard.edu/cancer/radiation-risk-from-medical-imaging

Rifkin, Benjamin, Michael Ackerman, and Judith Folkenberg. *Human Anatomy: Depicting the Body from the Renaissance to Today.* London: Thames and Hudson, 2006.

Röntgen, William. "Ueber eine neue Art von Strahlen. (On a New Kind of Rays.)" In *Classics of Orthopaedics*. Edited by Edgar Bick, 278–84. Philadelphia: Lippincott, 1976.

"Safety for Security Screening Using Devices That Expose Individuals to Ionizing Radiation." Accessed September 25, 2019. http://hps.org/publicinformation/ate/faqs/backscatterfaq.html

Sanders, Mark. "Historical Perspective: William Cheselden: Anatomist, Surgeon, and Medical Illustrator." *Spine* 24, no. 21 (1999): 2282–89.

Schultz, Kathryn, and Jennifer Wolf. "Emerging Technologies in Osteoporosis Diagnosis." *Journal of Hand Surgery, American* 44, no. 3 (2019): 240–43.

Shin, Eon, and Roy Meals. "The Historical Importance of the Hand in Advancing the Study of Human Anatomy." *Journal of Hand Surgery, American* 30, no. 2 (2005): 209–21.

Tehranzadeh, Jamshid. *Basic Musculoskeletal Imaging.* New York: McGraw-Hill Education, 2013.

Thomas, K. Bryn. "The Great Anatomical Atlases." *Proceedings of the Royal Society of Medicine* 67, no. 3 (1974): 223–32.

Webb, W. Richard, William Brant, and Nancy Major. *Fundamentals of Body CT.* 5th ed. Philadelphia: Elsevier, 2019.

Woodward, Paula. *Imaging Anatomy Ultrasound.* 2nd ed. Philadelphia: Elsevier, 2018.

Xing, Lida, Michael Caldwell, Rui Chen, Randall Nydam, Alessandro Palci, Tiago Simoes, and Ryan McLellar. "A Mid-Cretaceous Embryonic-to-Neonate Snake in Amber from Myanmar." *Science Advances* 4, no. 7 (2018): eaat5042.

## *Chapter 9:* THE FUTURE OF CONCEALED BONE

Antoniac, Julian, ed. *Bioceramics and Biocomposites: From Research to Clinical Practice.* Hoboken: John Wiley and Sons, 2019.

Ding, Zhen, Chao Yuan, Xirui Peng, Tiejun Wang, Jerry Qu, and Martin Dunn. "Direct 4D Printing via Active Composite Materials." *Science Advances* 3, no. 4 (2017): e1602890.

Inimuddin, Abdullah Asiri, and Ali Mohammad, eds. *Applications of Nanocomposite Materials in Orthopedics.* Duxford, UK: Woodhead Printing, 2019.

Kang, Hyun-Wook, Sang Jin Lee, In Kap Ko, Carlow Kengla, James Yoo, and Anthony Atala. "A 3D Bioprinting System to Produce Human-Scale Tissue Constructs with Structural Integrity." *Nature Biotechnology* 34, no. 3 (2016): 312–19.

Li, Bingyun, and Thomas Webster, eds. *Orthopedic Biomaterials: Progress in Biology, Manufacturing, and Industry Perspectives*. Cham, Switzerland: Springer, 2018.

Liu, Huinan, ed. *Nanocomposites for Musculoskeletal Tissue Regeneration*. Duxford, UK: Woodhead Publishing, 2016.

Maniruzzaman, Mohammed, ed. *3D and 4D Printing in Biomedical Applications: Process Engineering and Additive Manufacturing*. Weinheim, Germany: Wiley-VCH, 2019.

Meals, Roy. "A Vision of Hand Surgery over the Next 25 Years." *Journal of Hand Surgery, American* 26, no. 1 (2001): 3–7.

Scudera, Giles, and Alfred Tria, eds. *Minimally Invasive Surgery in Orthopedics*. 2nd ed. Berlin: Springer Verlag, 2019.

Zheng, Guoyan, Wei Tian, and Xiahai Zhuang, eds. *Intelligent Orthopaedics: Artificial Intelligence and Smart Image-Guided Technology for Orthopaedics*. Singapore: Springer, 2018.

### *Chapter 10:* BONE LEFT ALONE

Arnaud, G., S. Arnaud, A. Ascenzia, E. Bonucci, and G. Graziani. "On the Problem of Preservation of Human Bone in Sea-Water." *Journal of Human Evolution* 7, no. 5 (1978): 409–14.

Bennike, Pia. "The Early Neolithic Danish Bog Finds: A Strange Group of People!" In *Bog Bodies, Sacred Sites and Wetland Archaeology*, 27–32. Edited by Bryony Coles, John Coles, and Mogens Jorgensen. Exeter, UK: University of Exeter, 1999.

Briggs, C. S. "Did They Fall or Were They Pushed? Some Unresolved Questions about Bog Bodies." In *Bog Bodies: New Discoveries and New Perspectives*, 168–82. Edited by R. C. Turner and R. G. Scaife. London: British Museum Press, 1995.

Callaway, Ewen. "Skeleton Plundered from Mexican Cave Was One of the Americas' Oldest." *Nature* 549, no. 7670 (2017): 14–15.

Capasso, Luigi. "Herculaneum Victims of the Volcanic Eruptions of Vesuvius in 79 AD." *Lancet* 356, no. 9238 (2000): 1344–46.

Chamberlain, Andrew, and Michael Pearson. *Earthly Remains, The History and Science of Preserved Human Bodies*. London: British Museum Press, 2001.

Chatters, James, Douglas Kennett, Yemane Asmerom, Brian Kemp, Victor Polyak, Alberto Blank, Patricia Beddows, et al. "Late Pleistocene Human Skeleton and mtDNA Link Paleoamericans and Modern Native Americans." *Science* 344, no. 6185 (2014): 750–54.

Fischer, Christian. "Bog Bodies of Denmark and North-West Europe." In *Mummies, Disease & Ancient Cultures*, 237–62. 2nd ed. Edited by Aidan Cockburn,

Eve Cockburn, and Theodore Reyman. Cambridge, UK: Cambridge University Press, 1998.

Hodges, Glen. "Most Complete Ice Age Skeleton Helps Solve Mystery of First Americans." *National Geographic*. Accessed September 22, 2019. https://www.nationalgeographic.com/news/2014/5/140515-skeleton-ice-age-mexico-cave-hoyo-negro-archaeology/

Kappelman, John, Richard Ketcham, Stephen Pearce, Lawrence Todd, Wiley Akins, Matthew Colbert, Mulugeta Feseha, Jessica Maisano, and Adrienne Witzel. "Perimortem Fractures in Lucy Suggest Mortality from Fall Out of Tree." *Nature* 537, no. 7621 (2016): 503–507.

Lahr, M. Mirazon, F. Rivera, R. Power, A. Mounier, B. Copsey, F. Crivellaro, et al. "Inter-Group Violence among Early Holocene Hunter-Gatherers of West Turkana, Kenya." *Nature* 529, no. 7586 (2016): 394–98.

Lanham, Url. *The Bone Hunters: The Heroic Age of Paleontology in the American West*. Mineola, NY: Dover, 2011.

LePage, Michael. "Bird Caught in Amber 100 Million Years Ago Is Best Ever Found." New Scientist. Accessed September 22, 2019. https://www.newscientist.com/article/2133981-bird-caught-in-amber-100-million-years-ago-is-best-ever-found/

Levine, Joshua. "Europe's Famed Bog Bodies Are Starting to Reveal Their Secrets." *Smithsonian Magazine*. Accessed September 22, 2019. https://www.smithsonianmag.com/science-nature/europe-bog-bodies-reveal-secrets-180962770/

Lyman, R. Lee. *Vertebrate Taphonomy*. Cambridge, UK: Cambridge University Press, 1994.

Mastrolorenzo, Guiseppe, Pier Petrone, Mario Pagano, Alberto Incoronato, Peter Baxter, Antonio Canzanella, and Luciano Fattore. "Herculaneum Victims of Vesuvius in AD 79." *Nature* 410 (2001): 769–70.

Petrone, Pierpaolo, Piero Pucci, Alessandro Vergara, Angela Amoresano, Leila Birolo, Francesca Pane, Francesco Sirano, et al. "A Hypothesis of Sudden Body Fluid Vaporization in the 79 AD Victims of Vesuvius." *PLoS One* 13, no. 9 (2018): e0203210, 1–27.

Pickering, Travis, and Kristian Carlson. "Baboon Taphonomy and Its Relevance to the Investigation of Large Felid Involvement in Human Forensic Cases." *Forensic Science International* 11, no. 1 (2004): 37–44.

Raymunt, Monica. "Down on the Body Farm: Inside the Dirty World of Forensic Science." *The Atlantic*. Accessed September 22, 2019. https://www.theatlantic.com/technology/archive/2010/12/down-on-the-body-farm-inside-the-dirty-world-of-forensic-science/67241/

Ritche, Carson. *Bone and Horn Carving, A Pictorial History*. South Brunswick, NJ: A. S. Barnes, 1975.

Roberts, David. *Limits of the Known*. New York: W. W. Norton, 2018.

Sarvesvaran, R., and Bernard Knight. "The Examination of Skeletal Remains." *Malaysian Journal of Pathology* 16, no. 2 (1994): 117–26.

Sherratt, Emma, Maria Castaneda, Russell Garwood, D. Luke Mahler, Thomas Sanger, Anthony Herrel, Kevin de Queiroz, and Jonathan Losos. "Amber Fossils Demonstrate Deep-Time Stability of Caribbean Lizard Communities." *Proceedings of the National Academy of Sciences USA* 112, no. 32 (2015): 9961–66.

University of Heidelberg. "Human Bones in South Mexico: Stalagmite Reveals Their Age as 13,000 Years Old: Researchers Date Prehistoric Skeleton Found in Cave in Yucatán." Science Daily. Accessed September 22, 2019. https://www.sciencedaily.com/releases/2017/08/170831131259.htm

Wilford, John. "Mammal Bones Found in Amber for First Time." *New York Times*. Accessed September 22, 2019. https://www.nytimes.com/1996/04/16/science/mammal-bones-found-in-amber-for-first-time.html

Xing, Lida, Michael Caldwell, Rui Chen, Randall Nydam, Alessandro Palci, Tiago Simoes, Ryan McKellar, et al. "A mid-Cretaceous Embryonic-to-Neonate Snake in Amber from Myanmar." *Science Advances* 4, no. 7 (2018): eaat 5042, 1–8.

Xing, Lida, Edward Stanley, Bai Ming, and David Blackburn. "The Earliest Direct Evidence of Frogs in Wet Tropical Forests from Cretaceous Burmese Amber." *Scientific Reports* 8, no. 8770 (2018): 1–8.

## *Chapter 11:* DEFERENCE TO BONE

Brenner, Erich. "Human Body Preservation—Old and New Techniques." *Journal of Anatomy* 2014, no. 3 (2014): 316–44.

Chinese Buddhist Encyclopedia. "The Practices and Rituals of Tibetan Kapala Skull Caps." Accessed September 22, 2019. http://www.chinabuddhismencyclopedia.com/en/index.php?title=The_practices_and_rituals_of_Tibetan_Kapala_skull_caps

Chou, Hung-Hsiang. *Oracle Bone Collections in the United States*. Berkeley: University of California Press, 1976.

clutterbuck12. "Wesley Figures See the Light!" Accessed September 22, 2019. https://rylandscollections.wordpress.com/2014/10/13/wesley-figures-see-the-light/

Dean, Carolina. "Traditional Bone Reading with Chicken Bones." Carolina Conjure. Accessed September 22, 2019. https://www.carolinaconjure.com/traditional-bone-reading.html

Dhwty. "The Origins of Voodoo, the Misunderstood Religion." Ancient Origins. Accessed September 22, 2019. https://www.ancient-origins.net/history-ancient-traditions/origins-voodoo-misunderstood-religion-002933

Dibble, Harold, Vera Aldeias, Paul Goldberg, Shannon McPherron, Dennis Sandgathe, and Teresa Steele. "A Critical Look at Evidence from La Chapelle-aux-Saints Supporting an Intentional Neanderthal Burial." *Journal of Archaeological Science* 53, no. 1 (2015): 649–57.

Doughty, Caitlin. *From Here to Eternity. Traveling the World to Find a Good Death*. New York: W. W. Norton, 2017.

entheology.org. "The Taino World." Accessed September 22, 2019. http://www.entheology.org/edoto/anmviewer.asp?a=140

Ferlisi, Ani. "Bone Deep with Meaning: History and Symbolism of the Calvera." Accessed September 22, 2019. https://blog.alexandani.com/history-and-symbolism-of-the-calavera/

Gaudette, Emily. "What Is the Day of the Dead? How to Celebrate Dia de los Muertos without Being Offensive." *Newsweek*. Accessed September 22, 2019. https://www.newsweek.com/day-dead-dia-de-los-muertos-sugar-skulls-696811

Handa, O. C. *Buddhist Monasteries of Himachal*. New Delhi: Indus Publishing Company, 2005.

Hessler, Peter. *Oracle Bones: A Journey between China's Past and Present*. New York: HarperCollins, 2006.

Hunt, Katie. "Hanging Coffins: China's Mysterious Sky Graveyards." CNN. Accessed September 22, 2019. https://www.cnn.com/travel/article/china-hanging-coffins/index.html

Johnston, Franklin. *The Lost Field Notes of Franklin R. Johnston's Life and Work Among the American Indians*. St. Louis: First Glance Books, 1997.

Koudounaris, Paul. *The Empire of Death. A Cultural History of Ossuaries and Charnel Houses*. London: Thames and Hudson, 2011.

Koudounaris, Paul. *Heavenly Bodies. Cult Treasures and Spectacular Saints from the Catacombs*. London: Thames and Hudson, 2013.

Koudounaris, Paul. *Memento Mori. The Dead among Us*. London: Thames and Hudson, 2015.

Lasseteria. "Pointing the Bone." Accessed September 22, 2019. http://www.lasseteria.com/CYCLOPEDIA/215.htm

Lieberman, Philip. *Uniquely Human: The Evolution of Speech, Thought, and Selfless Behavior*. Cambridge, MA: Harvard University Press, 1991.

Lipke, Ian. "Curses and Cures: Superstitions." Unusual Historicals. Accessed September 22, 2019. http://unusualhistoricals.blogspot.com/2014/11/curses-and-cures-superstitions.html

Loseries-Leick, Andrea. *Tibetan Mahayoga Tantra: An Ethno-Historical Study of Skulls, Bones, and Relics*. Dehli: B. R. Publishing, 2008.

Madison, Paige. "Who First Buried the Dead?" Aeon. Accessed September 22, 2019. https://aeon.co/essays/why-we-should-bury-the-idea-that-human-rituals-are-unique

Metropolitan Museum of Art. "Rkangling." Accessed September 22, 2019. https://www.metmuseum.org/art/collection/search/505032?&searchField=All&sortBy=Relevance&ft=bone+trumpet&offset=0&rpp=80&pos=24

Murphy, Eileen, ed. *Deviant Burial in the Archaeological Record*. Oxford, UK: Oxbow Press, 2008.

mysafetysign.com. "History of the Skull & Crossbones and Poison Symbol." Accessed September 22, 2019. https://www.mysafetysign.com/poison-symbol-history

NaNations. "Tree and Scaffold Burial." Accessed September 22, 2019. http://www.nanations.com/burialcustoms/scaffold_burial.htm

Romey, Kristin. "Ancient Shark Fishermen Found Buried with Extra Limbs." *National Geographic*. Accessed September 22, 2019. https://www.nationalgeographic.com/news/2018/04/peru-viru-ancient-shark-fishermen-archaeology/

Shafik, Vervat, Ashraf Selim, Isam Seikh, and Zahi Hawass. "Computed Tomography of King Tut-Ankh-Amen." The Ambassadors. Accessed September 22, 2019. https://ambassadors.net/archives/issue23/selectedstudy3.htm

Spiegel. "Roll Over Dracula: 'Vampire Cemetery' Found in Poland." ABC News. Accessed September 22, 2019. https://abcnews.go.com/International/roll-dracula-vampire-cemetery-found-poland/story?id=19739673

Surname Database. "Last Name: Brisbane." Accessed September 22, 2019. http://www.surnamedb.com/Surname/Brisbane

Taino Museum. "Double Vomiting Stick Made of Bone." Accessed September 22, 2019. https://tainomuseum.org/portfolio-view/double-vomiting-stick-made-bone/

Tayanin, Damrong. "Divination by Chicken Bones. A Tradition among the Kammu in Northern Lao People's Democratic Republic." Accessed September 22, 2019. https://person2.sol.lu.se/DamrongTayanin/divination.html

Trimble, Marshall. "An Old Photograph Depicts an Indian Burial Scaffold with a Dead Horse in the Foreground. Was That Normal?" True West. Accessed September 21, 2019. https://truewestmagazine.com/an-old-photograph-depicts-an-indian-burial-scaffold-with-a-dead-horse-in-the-foreground-was-that-normal/

University of Cambridge. "World First as 3,000-Year-Old Chinese Oracle Bones Go 3D." Accessed September 22, 2019. https://www.cam.ac.uk/research/news/world-first-as-3000-year-old-chinese-oracle-bones-go-3d

Vatican. "Catacombs of Rome." Accessed September 22, 2019. http://www.vatican.va/roman_curia/pontifical_commissions/archeo/inglese/documents/rc_com_archeo_doc_20011010_catacroma_en.html

wikipedia.com. "Totenkopf." Accessed September 22, 2019. https://en.wikipedia.org/wiki/Totenkopf

Zimmerman, Fritz. "Native American Burials: Trees and Scaffolds Illustrated." Accessed September 22, 2019. https://americanindianshistory.blogspot.com/2011/07/native-american-burials-trees-and.html

## *Chapter 12:* BONES THAT TEACH

Alden, Andrew. "Potassium-Argon Dating Methods." Accessed September 22, 2019. https://www.thoughtco.com/potassium-argon-dating-methods-1440803

Bahn, Paul, ed. *The Archaeology Detectives*. Pleasantville, NY: Reader's Digest, 2001.

Bello, Silvia, Rosalind Wallduck, Simon Parfitt, and Chris Stringer. "An Upper Palaeolithic Engraved Human Bone Associated with Ritualistic Cannibalism." *PLoS One* 12, no. 8 (2017): e0182127, 1–18.

Bryson, Bill. *A Short History of Nearly Everything.* New York: Broadway Books, 2003.

Dirkmaat, Dennis, and Luis L. Cabo. "Forensic Anthropology: Embracing the New Paradigm." In *A Companion to Forensic Anthropology*, 3–40. Edited by Dennis Dirkmaat. Malden, MA: Wiley-Blackwell, 2012.

Gibbons, Ann. "The Human Family's Earliest Ancestors." *Smithsonian Magazine.* Accessed September 23, 2019. https://www.smithsonianmag.com/science-nature/the-human-familys-earliest-ancestors-7372974/

Goodrum, Matthew, and Cora Olson. "The Quest for an Absolute Chronology in Human Prehistory: Anthropologists, Chemists and the Fluorine Dating Method in Paleoanthropology." *British Journal of the History of Science* 42, no. 1 (2009): 95–114.

Gould, Stephen. *The Mismeasure of Man.* New York: W. W. Norton, 1996.

Gresky, Julia, Juliane Haelm, and Lee Clare. "Modified Human Crania from Göbekli Tepe Provide Evidence for a New Form of Neolithic Skull Cult." *Science Advances* 3, no. 6 (2017): e1700564, 1–10.

Harrison, Simon. "Bones in the Rebel Lady's Boudoir: Ethnology, Race and Trophy-Hunting in the American Civil War." *Journal of Material Culture* 15, no. 4 (2010): 385–401.

Haslam, Michael, ed. *Archaeological Science Under a Microscope. Studies in Ancient Residue and Ancient DNA Analysis in Honour of Thomas H. Loy.* Canberra: ANU Press, 2009.

Henke, Winfried, and Ian Tattersall. *Handbook of Paleoanthropology.* Berlin: Springer-Verlag, 2007.

Hirst, Kris. "Archaeological Dating: Stratigraphy and Seriation." Accessed September 22, 2019. https://www.thoughtco.com/archaeological-dating-stratigraphy-and-seriation-167119

Hirst, Kris. "Midden: An Archaeological Garbage Dump." Accessed September 22, 2019. https://www.thoughtco.com/midden-an-archaeological-garbage-dump-171806

Kappelman, John, Richard Ketcham, Stephen Pearce, Lawrence Todd, Wiley Akins, Matthew Colbert, Mulugeta Feseha, Jessica Maisano, and Adrienne Witzel. "Perimortum Fractures in Lucy Suggest Mortality from Fall Out of a Tree." *Nature* 537, no. 7621 (2016): 503–7.

Kilgrove, Kristina. "Is That Skeleton Gay? The Problem With Projecting Modern Ideas onto the Past." Forbes. Accessed September 23, 2019. https://www.forbes.com/sites/kristinakillgrove/2017/04/08/is-that-skeleton-gay-the-problem-with-projecting-modern-ideas-onto-the-past/#598db1ef30e7

Lanham, Uri. *The Bone Hunters.* New York: Columbia University Press, 1973.

Mays, Simon. *The Archaeology of Human Bones.* 2nd ed. London: Routledge, 2010.

McNish, James. "Carved Bone Reveals Rituals of Prehistoric Cannibals." Natural History Museum. Accessed September 22, 2019. https://www.nhm.ac.uk/discover/news/2017/august/carved-bone-reveals-rituals-of-prehistoric-cannibals.html

Meyer, Christian, Christian Lohr, Detlef Gronenborn, and Kurt Alt. "The Massacre Mass Grave of Schöneck-Kilianstädten Reveals New Insights into Collective Violence in Early Neolithic Central Europe." *Proceedings of the National Academy of Sciences USA* 112, no. 36 (2015): 11217–22.

Price, Michael. "Study Reveals Culprit Behind Piltdown Man, One of Science's Most Famous Hoaxes." Science Magazine. Accessed September 23, 2019. https://www.sciencemag.org/news/2016/08/study-reveals-culprit-behind-piltdown-man-one-science-s-most-famous-hoaxes

Price, T. Douglas, Robert Frei, Ute Brinker, Gundula Lidke, Thomas Terberger, Karin Frei, and Detlef Jantzen. "Multi-Isotope Proviencing of Human Remains from a Bronze Age Battlefield in the Tollense Valley in Northeast Germany." *Archaeological and Anthropological Sciences* 11, no. 1 (2019): 33–49.

Pyne, Lydia. *Seven Skeletons. The Evolution of the World's Most Famous Human Fossils*. New York: Viking, 2016.

Redman, Samuel. *Bone Rooms: From Scientific Racism to Human Prehistory in Museums*. Cambridge, MA: Harvard University Press, 2016.

Richter, Daniel, Rainer Gruen, Renaud Joannes-Boyau, Teresa Steel, Fethi Amani, Mathiew Rue, Paul Fernandes, et al. "The Age of the Hominin Fossils from Jebel Irhoud, Morocco, and the Origins of the Middle Stone Age." *Nature* 546 (2017): 293–96.

Russell, Miles. *The Piltdown Man Hoax. Case Closed*. Stroud, UK: History Press, 2012.

Shorto, Russell. *Descartes' Bones. A Skeletal History of the Conflict Between Faith and Reason*. New York: Vintage, 2008.

Swisher, C. III, Garniss Curtis, and Roger Lewin. *How Two Geologists' Dramatic Discoveries Changed Our Understanding of the Evolutionary Path to Modern Humans*. New York: Scribner, 2000.

Trinkhaus, Erik, and Pat Shipman. *The Neanterthals: Changing the Image of Mankind*. New York: Knopf, 1993.

UC Museum of Paleontology. "Othneil Charles Marsh." Accessed September 22, 2019. https://ucmp.berkeley.edu/history/marsh.html

Von Koenigswald, Gustav. *Meeting Prehistoric Man*. Translated by Michael Bullock. New York: Harper, 1956.

Walker, Alan, and Pat Shipman. *The Wisdom of the Bones, in Search of Human Origins*. New York: Vintage, 1997.

Wesch, Michael. *The Art of Being Human: A Textbook for Cultural Anthropology*. Manhattan, KS: New Prairie Press, 2018.

Winchester, Simon. *Skulls. An Exploration of Alan Dudley's Curious Collection*. New York: Black Dog and Leventhal, 2012.

Zupancich, Andrea, Stella Nunziante-Cesaro, Ruth Blasco, Jordi Rosell, Emanuella Cristiani, Flava Vendetti, Cristina Lemorini, Ran Barkai, and Avi Gopher. "Early Evidence of Stone Tool Use in Bone Working Activities at Qesem Cave, Israel." *Scientific Reports* 6, no. 37686 (2016): 1–7.

## *Chapter 13:* THE BUSINESS OF BONES

Barnett, LeRoy. "How Buffalo Bones Became Big Business." *North Dakota History* 39, no. 1 (1972): 20–24.

Ewers, John C. "Hair Pipes in Plains Indian Adornment: A Study in Indian and White Ingenuity." *Bulletin / Smithsonian Institution, Bureau of American Ethnology* no. 164. Anthropological Papers no. 50 (1957): 29–85.

Frugoni, Chiara. *Books, Banks, Buttons, and Other Inventions from the Middle Ages*. New York: Columbia University Press, 2003.

Lessem, Don. "Don't Believe the Anti-Government Tale Spun by This New Dinosaur Documentary." Slate. Accessed September 23, 2019. https://slate.com/culture/2014/08/dinosaur-13-review-movie-about-peter-larson-spins-a-bogus-tale.html

"Minot North Dakota and the Buffalo Bone Trade." *North Dakota History* 39, no. 1 (1972): 23–42.

Mould, Quita, Ian Carliske, and Esther Cameron. *Craft, Industry and Everyday Life: Leather and Leatherworking in Anglo-Scandinavian and Medieval York*. Micklegate, UK: York Archaeological Trust, 2004.

Rare Historical Photos. "Bison Skulls to Be Used for Fertilizer, 1870." Accessed September 25, 2019. https://rarehistoricalphotos.com/bison-skulls-pile-used-fertilizer-1870/

Ritche, Carson. *Bone and Horn Carving, A Pictorial History*. South Brunswick, NJ: A. S. Barnes, 1975.

Smith, Stacy Vanek, host. "Planet Money, Episode 660: The T-Rex in My Backyard." NPR. Accessed September 23, 2019. https://www.npr.org/sections/money/2015/10/30/453257199/the-t-rex-in-my-backyard

Tomasi, Michele. *La Botegga degli Embriachi*. Florence, Italy: The National Museum of the Bargello, 2001.

Williamson, Paul. *Medieval Ivory Carvings: 1200–1550*. London: V & A Publishing, 2014.

## *Chapter 14:* DOMESTIC BONES

Bahn, Paul, ed. *The Archaeology Detectives*. Pleasantville, NY: Reader's Digest, 2001.

Bandi, Hans-Georg. "A Yupiget (St. Lawrence Island Yupik) Figurine as a Historical Record." *Alaska Journal Anthropology* 4, no. 1–2 (2006): 148–54.

Bunn, Henry, and Alia Gurtov. "Prey Mortality Profiles Indicate That Early Pleistocene Homo at Olduvai Was an Ambush Predator." *Quaternary International* 322–323 (2014): 44–53.

Corbett, Debra. "Two Chiefs' Houses from the Western Aleutian Islands." *Arctic Anthropology* 48, no. 2 (2011): 3–16.

Dawson, Peter. "Interpreting Variability in Thule Inuit Architecture: A Case Study from the Canadian High Arctic." *American Antiquity* 66, no. 3 (2001): 453–70.

Dominy, Nathaniel, Samuel Mills, Christopher Yakacki, Paul Roscoe, and Dana Carpenter. "New Guinea Bone Daggers Were Engineered to Preserve Social Prestige." *Royal Society Open Science* 5, no. 172067 (2018): 1–12.

Ferraro, Joseph, Thomas Plummer, Briana Pobiner, James Oliver, Laura Bishop, David Braun, Peter Ditchfield, et al. "Earliest Archaeological Evidence of Persistent Hominin Carnivory." *PLoS One* 8, no. 4 (2013): e62174, 1–10.

Geggel, Laura. "Iron Age People in Scotland Really Knew How to Party, Ancient Trash Heap Reveals." Live Science. Accessed September 23, 2019. https://www.livescience.com/62138-iron-age-meat-feast-with-jewelry.html

Hirst, Kris. "Arctic Architecture—Paleo-Eskimo and Neo-Eskimo Houses." Accessed September 23, 2019. https://www.thoughtco.com/paleo-and-neo-eskimo-houses-169871?utm_source=pinterest&utm_medium=social&utm_campaign=mobilesharebutton2

Hirst, Kris. "Midden: An Archaeological Garbage Dump." Accessed September 23, 2019. https://www.thoughtco.com/midden-an-archaeological-garbage-dump-171806

Jeater, Meriel. "How Did Medieval Londoners Celebrate Christmas?" Museum of London. Accessed September 23, 2019. https://www.museumoflondon.org.uk/discover/how-did-medieval-londoners-celebrate-christmas

Jones, Fancesca, Lauren Gilmour, and Martin Henig. *Treasures of Oxfordshire.* Oxford, UK: Friends of Archives, Museums and Oxfordshire Studies, 2004.

Klopfer, J. E. "The Nutmeg Grater: A Kitchen Collectible, and So Much More." *Journal of Antiques and Collectibles.* Accessed September 23, 2019. http://journalofantiques.com/features/nutmeg-grater-kitchen-collectible-much/

Lowe, Stephanie. "The World's Oldest Building: The Fossil Cabin at Como Bluff." Accessed September 23, 2019. https://www.wyohistory.org/encyclopedia/fossil-cabin

MacGregor, Arthur. *Bone, Antler, Ivory and Horn. The Technology of Skeletal Materials Since the Roman Period.* New York: Routledge, 2015.

MacGregor, Elizabeth. *Craft, Industry and Everyday Life: Bone, Antler, Ivory and Horn from Anglo-Scandinavian and Medieval York.* Micklegate, UK: Council for British Archaeology, 1999.

Magnusson, Halldor. "Cannon Bones: The Dark Age Boneworker's Best Source." Halldor the Viking. Accessed September 23, 2019. https://halldorviking.wordpress.com/2013/04/03/cannon-bones-the-dark-age-boneworkers-best-resource/

McLagan, Jennifer. *Bones. Recipes, History, and Lore.* New York: William Morrow, 2005.

Nelson, Edward. *The Eskimo about Bering Strait.* Washington, DC: Government Printing Office, 1900.

North, S. N. Dexter. "The Development of American Industries since Columbus.

V. The Manufacture of Wool." *Popular Science Monthly* 39 (May–October 1891): 176–95.

Office of the State Archaeologist. "Bone Tools." University of Iowa. Accessed September 23, 2019. https://archaeology.uiowa.edu/bone-tools-0

Rhodes, Michael. "A Pair of Fifteenth-Century Spectacle Frames from the City of London." *Antiquaries Journal* 62, no. 1 (1982): 57–73.

Roberts, Phil. "The Builder of the 'World's Oldest Cabin.'" University of Wyoming. Accessed September 23, 2019. https://web.archive.org/web/20090427155026/http://uwacadweb.uwyo.edu/ROBERTSHISTORY/worlds_oldest_cabin_fossil.htm

Schwatka, Frederick. "The Igloo of the Inuit.—III." *Science* 2, no. 30 (1883): 259–62.

## *Chapter 15:* BEGUILING BONES

Gardner, Jane. *Henry Moore. From Bones and Stones to Sketches and Sculptures*. New York: Four Winds Press, 1993.

Gray, Henry. *Anatomy of the Human Body*. 20th ed. Edited by Warren Lewis. Philadelphia: Lea and Febiger, 1918.

Henry Moore Foundation. "Biography." Accessed September 25, 2019. https://www.henry-moore.org/about-henry-moore/biography

Jansen, Jan, and Wouter van Gestel. "Cleaning Skulls and Skeletons by Maceration." Accessed September 23, 2019. https://skullsite.com/skull-cleaning-tutorial/

Mortensen, Jenna. "Astragaloi: Greco-Roman Dice Oracles." Accessed September 23, 2019. https://ladyofbones.files.wordpress.com/2013/06/astragaloi-handout.pdf

Museum of London. "Bone Skates: 12th Century." Accessed December 2, 2019. https://www.museumoflondonprints.com/image/61275/bone-skates-12th-century

Neves, Rogerio, Gregory Saggers, and Ernest Manders. "Lizard's Leg and Howlet's Wing: Laboratory Preparation of Skeletal Specimens." *Plastic and Reconstructive Surgery* 96, no. 4 (1995): 992–94.

Ritche, Carson. *Bone and Horn Carving, A Pictorial History*. South Brunswick, NJ: A. S. Barnes, 1975.

Scott, Heather. "Understanding Bow Tip Plates." Strings. Accessed September 23, 2019. http://stringsmagazine.com/understanding-bow-tip-plates/

Spitzers, Thomas. "Die Konstanzer Paternosterleisten—Analyse zur Technik und Wirtschaft im spätmittelalterlichen Handwerk der Knochenperlenbohrer." *Findings from Baden-Württemberg* 33 (2013): 661–940.

Verrill, A. Hyatt. *The Real Story of the Whaler: Whaling, Past and Present*. New York: Appleton, 1923.

### *Chapter 16:* THE FUTURE OF REVEALED BONE

Cunningham, John, Imran Rahman, Stephan Lautenschlaager, Emily Rayfield, and Philip Donoghue. "A Virtual World of Paleontology." *Trends in Ecology and Evolution* 29, no. 6 (2014): 347–57.

Fages, Antoine, Kristian Hanghøj, Naveet Khan, Charleen Gaunitz, Andaine Seguin-Orlando, Micheela Leonardi, Christian Constanz, et al. "Tracking Five Millennia of Horse Management with Extensive Ancient Genome Time Series." *Cell* 177, no. 6 (2019): 1419–35. e31.

Geggel, Laura. "Mammoth DNA Briefly 'Woke Up' Inside Mouse Eggs. But Cloning Mammoths Is Still a Pipe Dream." Live Science. Accessed September 26, 2019. https://www.livescience.com/64998-mammoth-cells-inserted-in-mouse-eggs.html

Hanson, Joe. "700,000-Year-Old Horse Genome Shatters Record for Sequencing of Ancient DNA." Wired. Accessed September 26, 2019. https://www.wired.com/2013/06/ancient-horse-genome/

Haslam, Michael, ed. *Archaeological Science under a Microscope: Studies in Ancient Residue and Ancient DNA Analysis in Honour of Thomas H. Loy*. Canberra: ANU Press, 2009.

Heintzman, Peter, Grant Zazula, Ross MacPhee, Eric Scott, James Cahill, Brianna McHorse, Joshua Kapp, et al. "A New Genus of Horse from Pleistocene North America." *eLife* 6 (2017): e29944.

Henke, Winfried, and Ian Tattersall. *Handbook of Paleoanthropology*. Berlin: Springer-Verlag, 2007.

Leake, Jonathan: "Science Close to Creating A Mammoth." The Times. Accessed September 26, 2019. https://www.thetimes.co.uk/article/science-close-to-creating-a-mammoth-z8zlvbgr9fl

Plotnick, Roy. "Beyond the Hammer and Whisk Broom: The Technology of Paleontology." Accessed September 23, 2019. https://medium.com/@plotnick/beyond-the-hammer-and-whisk-broom-the-technology-of-paleontology-c81088e2164d

Presslee, Samantha, Graham J. Slater, François Pujos, Analía M. Forasiepi, Roman Fischer, Kelly Molloy, Meaghan Mackie, et al. "Palaeoproteomics Resolves Sloth Relationships." *Nature Ecology and Evolution* 3, no. 7 (2019): 1121–30.

Yamagata, Kazuo, Kouhei Nagai, Hiroshi Miyamoto, Masayuki Anzai, Hiromi Kato, Key Miyamoto, Satoshi Kurosaka, et al. "Signs of Biological Activities of 28,000-Year-Old Mammoth Nuclei in Mouse Oocytes Visualized by Live-cell Imaging." *Scientific Reports* 9, no. 4050 (2019): 1–12.

# *Index*

Page numbers in *italic* refer to figures.